高等职业院校精品教材系列

市政工程专业英语
（道路与桥梁方向）

李　丽　郭春明　主编

電子工業出版社
Publishing House of Electronics Industry
北京 · BEIJING

内 容 简 介

本书是根据教育部新的职业教育教学改革要求，以培养学生用英语作为工具交流市政工程技术信息的能力为主要目标，专门针对市政工程、道路与桥梁工程、交通工程等专业英语课程的教学目标进行编写的。全书内容分为4章，第1章为专业英语的特点及其翻译策略；第2章为土木工程专业基础，涉及建筑材料、道路测量技术、结构设计、标书与合同等；第3章为道路工程；第4章为桥梁工程。本书的每章设有教学导航、知识分布网络、知识梳理与总结、思考与练习题等，以方便教学。

本书为高等职业本专科院校市政工程、道路与桥梁工程、交通工程等专业英语课程的教材，也可作为开放大学、成人教育、自学考试、中职学校、培训班的教材，以及土木工程技术人员的参考书。

本教材配有免费的电子教学课件、习题参考答案、参考译文等，详见前言。

图书在版编目（CIP）数据

市政工程专业英语：道路与桥梁方向 / 李丽，郭春明主编. —北京：电子工业出版社，2023.12
高等职业院校精品教材系列
ISBN 978-7-121-36575-1

Ⅰ. ①市… Ⅱ. ①李… ②郭… Ⅲ. ①市政工程－英语－高等学校－教材 Ⅳ. ①TU99

中国版本图书馆 CIP 数据核字（2019）第 092812 号

责任编辑：陈健德（E-mail：chenjd@phei.com.cn）
印　　刷：北京虎彩文化传播有限公司
装　　订：北京虎彩文化传播有限公司
出版发行：电子工业出版社
　　　　　北京市海淀区万寿路 173 信箱　邮编　100036
开　　本：787×1 092　1/16　印张：10.5　字数：269 千字
版　　次：2023 年 12 月第 1 版
印　　次：2023 年 12 月第 1 次印刷
定　　价：45.00 元

凡所购买电子工业出版社图书有缺损问题，请向购买书店调换。若书店售缺，请与本社发行部联系，联系及邮购电话：（010）88254888，88258888。

质量投诉请发邮件至 zlts@phei.com.cn，盗版侵权举报请发邮件至 dbqq@phei.com.cn。

本书咨询联系方式：chenjd@phei.com.cn。

前言

按照教育部有关大学英语教学的规定，大学英语教学分为基础阶段和应用提高阶段。在基础阶段主要是公共英语教学，分为大学英语一至六级；在应用提高阶段包括专业英语和高级英语教学两部分。非英语专业的学生在完成基础阶段的学习任务后（通过国家四级、六级考试），需要继续学习专业英语。

本书是根据教育部新的职业教育教学改革要求，以培养学生用英语作为工具交流市政工程技术信息的能力为主要目标，专门针对市政工程、道路与桥梁工程、交通工程等专业英语课程的教学目标进行编写的。专业英语作为基础英语的后续课程，重点是培养学生阅读和翻译英文专业书刊的能力，提高阅读、翻译文献资料的质量和速度，了解国外同行业的发展和相关信息。通过对本书的学习，学生能够掌握必要的专业英语词汇，把英语学习与专业知识学习有机地结合在一起，培养科技英语的阅读、翻译和写作能力，提高以英语为工具获取专业所需信息的能力，为学生日后的工作、科学研究以及国际学术交流等打下良好的英语基础。

本书在内容选取时着重考虑其学术性、实用性、可读性及学生的接受程度等。全书内容分为 4 章，第 1 章为专业英语的特点及其翻译策略；第 2 章为土木工程专业基础，涉及建筑材料、道路测量技术、结构设计、标书与合同等；第 3 章为道路工程；第 4 章为桥梁工程。本书的参考学时为 42 学时，在使用时可根据实际教学情况适当调整。

本书为高等职业本专科院校市政工程、道路与桥梁工程、交通工程等专业英语课程的教材，也可作为开放大学、成人教育、自学考试、中职学校、培训班的教材，以及土木工程技术人员的参考书。

结合编者多年的专业英语教学经验，在保持以往同类教材优点的基础上，本书在内容选取、章节设置等方面具有以下特色：

（1）在内容选取上，既注重专业基础内容，又关注目前的研究热点，力争实现“引导性”“知识性”和“前瞻性”的有机结合。

（2）在章节设置上，设有 4 章 18 节。每章设有教学导航、知识分布网络、知识梳理与总结、思考与练习题，每节后均配有注释、生词、短语内容，以方便教学。

（3）本书增加许多图片，加强比较直观的教学效果，以期激发读者的阅读兴趣。

本书由黑龙江建筑职业技术学院李丽、郭春明任主编，由黑龙江建筑职业技术学院边喜龙、吕君主审。在本书的编写过程中，参考了相关的资源及书籍，在此对其作者表示衷心感谢。

由于编者水平所限，疏漏及不妥之处在所难免，敬请读者批评指正，使本书在使用的过程中不断得到改进。

为了方便教师教学，本书还配有免费的电子教学课件、习题参考答案、参考译文等，请有此需要的教师扫一扫书中二维码阅览或登录华信教育资源网（http://www.hxedu.com.cn）免费注册后进行下载，有问题时请在网站留言或与电子工业出版社联系（e-mail:hxedu@phei.com.cn）。

编者

扫一扫看本课程教学导航设计

Chapter 1　Characteristics of Professional English and its Translation Tactics

第 1 章　专业英语的特点及其翻译策略

教学导航

教	知识重点	1. Characteristics of Professional English 专业英语的特点； 2. Translation Tactics of Professional English 专业英语的翻译策略
	知识难点	Translation Tactics of Professional English 专业英语的翻译策略
	推荐教学方式	明确教学目标，选择适当的教学环节采用任务教学法，依据期刊论文摘要等内容，通过阅读、翻译等过程让学生熟悉市政工程等专业英语的特点及翻译策略
	建议学时	6 学时
学	推荐学习方法	以任务驱动和小组讨论的学习方式为主。结合本章内容，通过参考资料，理解市政工程等专业词汇和专业术语的学习方法，培养专业英语翻译能力
	必须掌握的理论知识	1. 专业英语的基本特点； 2. 学会运用适当的翻译方法和技巧，在忠实于原文的基础上，将原文准确地表达出来； 3. 英文摘要的写作
	必须掌握的技能	通过对本章的学习，学生能够系统地掌握市政工程等专业英语的文体特征和专业文献的阅读方法，掌握英语资料的阅读、翻译以及英文摘要写作的技巧

1.1 Characteristics of Professional English 专业英语的特点

扫一扫看本节教学课件

英语可分为普通英语、科技英语和专业英语等，具体区别见表 1-1。

表 1-1 专业英语与其他分支的关系与区别

英语 English	普通英语 General English (GE)	普通英语又称日常英语
	科技英语 English for Science and Technology (EST)	科技英语是英语的一种语体，包括自然科学和社会科学的学术著作、论文、研究报告、专利产品的说明等。科技英语注重科学性、逻辑性、正确性与严密性，在词汇、语法、修饰等方面具有自己的特色
	专业英语 English for Special Purposes (ESP)	科技英语的进一步专业化，即为专业英语。专业英语是随着新学科的不断涌现和专业分工的日益细化，在科技英语的基础上逐步形成的

专业英语与普通英语有相同的语言系统和语法规则，但也存在着明显差别。专业英语文章属于科技文章范畴，其中不仅有大量的专业词汇和专业术语，还有许多的合成新词和缩略词，但两者的主要区别在于文体差异。专业英语主要是对客观事实和客观真理进行论述，逻辑性强，条理通顺。另外专业英语的语法结构也有其自身的特性，如长句多，被动语态、非限定动词或限定定语从句的使用频率高等。由于专业英语与专业内容紧密配合，相互一致，懂专业的人用起来得心应手，而不懂专业的人用起来则困难重重。因而必须具有一定的相关专业知识基础，才能正确地理解和运用专业英语。对学习者而言，学习专业英语是为了了解专业英语的特点，掌握相关的英语基础知识，但更重要的是在学习专业英语（如文献阅读）的过程中积累和扩充专业知识，实际上，这也是学习专业英语的最终目的。专业英语和专业知识是密不可分的，如何将两者融会贯通，值得重视。

与普通英语和文学英语相比，专业英语主要有以下四个方面的特点。

1.1.1 语言特点

专业英语作为英语的一种应用文体，正在逐渐成为工程技术人员与国外同行进行信息交流的重要手段。对专业英语一般要求严谨周密、概念准确、逻辑性强、行文简练、重点突出、句式严整、少有变化，常用前置性陈述，即在句中尽量将主要信息前置，通过主语传递主要信息。

简而言之，专业英语的语言特点：清晰、准确、精练、严密。

【例 1】 Rigid pavements are made up of Portland cement concrete and may or may not have a base course between the pavement and the subgrade. It is different from flexible pavement which includes several layers of structural components of pavement. In the rigid pavement, the concrete, exclusive of the base, is referred to as the pavement.

译文： 刚性路面由硅酸盐水泥混凝土组成，在路面和路基之间可能有也可能没有基层。它与含有多层路面结构层的柔性路面不同。在刚性路面中，只有混凝土层被称为路面，不包括基层。

1.1.2　词汇特点

市政工程等专业的英语词汇与道路工程、桥梁工程、建筑材料、施工技术、结构设计以及工程管理等密切相关，有一定数量的专业词汇和术语。同时，专业英语词汇的内涵比较广泛，词的用法灵活，一词多义、一词多用等现象较多。加上大量复合词、缩略词以及各种构词法所创造的新词，使得专业英语词汇错综复杂且歧义颇多，需要从专业内容上去理解记忆。

例如：stress 在基础英语中通常被翻译成为“压力，强调，重音”，而在市政工程等专业英语中常以“应力”词义出现，如“stress intensity”应译为“应力强度”而不是“压力强度”；**weather** 除了“天气”，还有“风化、侵蚀”的意思，如“weathered rock”应译为“风化岩”而不能译为“天气岩”；**moment** 除了“片刻”，还有“力矩”的意思，比如“overturning moment”译为“倾覆力矩”而不译为“倾覆时刻”。其他的专业词汇如“civil engineering”译为“土木工程”而不译为“民用工程”；“钢束”（bundle reinforcement）不能按照字面意思直接翻译为“steel beam”等等。由此可见，专业英语翻译中词义的选择，必须符合专业术语翻译规范，从专业内容上去判断词义，才能准确、恰当地表达特定概念。

市政工程等专业有一定数量的专业词汇和术语。例如，对道路工程专业，有 pavement（路面）、roadbed（路基）、super-elevation（超高）、lateral clearance（侧向余宽）、grade change point（变坡点）、designed elevation（设计高程）等；对桥梁工程专业，有 abutment（桥台）、pier（桥墩）、deck（桥面）、caisson（沉井）、box girder（箱梁）、cofferdam（围堰）等。

一般说来，专业文献中的专业词汇（或科技词汇）有三类：

第一类是纯专业词汇。它的意义很单纯，只有一种专业含义。有时候则是根据需要而造出来的，如 T-beam（T 形梁）、fire-proof brick（耐火砖）、cable-stayed bridge（斜拉桥）等。

第二类是半专业词汇。半专业词汇大多在各个专业中通用，但在不同的专业领域可能有不同的含义，如 frame（框架、屋架、机座、体系、画面等）、operation（操作、运行、运算、作业、效果等）、load（负载、加载、装入、输入等）等。

第三类是非专业词汇。这类词汇在非专业英语中使用不多，但却严格属于非专业英语性质。这类词汇的数量很多，如 application（应用、用途、作用、申请等）、implementation（实现、执行、运行等）、yield（产生、得出、发出等）等。

专业英语较多地使用了合成法、派生法、转换法及缩略法等构词法，掌握专业英语词汇的构词法，有助于扩大词汇量。构词法的分类、定义及例词见表 1-2。

表 1-2　专业英语构词法的分类、定义及例词

构词法	定　义	例　词
合成法	把两个或两个以上具有独立意思的单词合成一个新词的方法叫作合成法	expressway（快速道路），highway（公路，道路），brickwork（砖圬工，砖结构），roadbed（路基），parkway（公园道路，风景区干道）
派生法	据有关专家统计，现代专业科技英语中，有 50%以上的词汇源于希腊语、拉丁语等外来语，而这些外来语词汇构成的一个主要特征就是大量使用词缀（前缀和后缀）和词根	multilane（多车道的），interchange（互通式立体交叉），frictional（摩擦的），geological（地质学的），sub-base（底基层）

续表

构词法	定　义	例　词
转换法	专业英语也较多地使用了词性转换方法。转换后词意往往与原来的词意相关。常见的词性转换类型有：名词→动词，形容词→动词，动词→名词，形容词→名词等	light（n.光，灯） →（v.点燃；adj.轻的，明亮的）， cut（v.切割）→（n.路堑，挖方）
缩略法	在阅读和撰写专业文献时，常常会遇到一些专有词汇或术语、物理量等的缩写，以及一些政府机构、学术团体、科技期刊和文献等的简称。缩略法包括如下四种形式： （1）省头；（2）省尾； （3）省头尾；（4）首字母缩写	phone（telephone）电话， Fig.（Figure）图， Eq.（Equation）方程（式）， in.（inch）英寸，flu（influenza）流感， QC（Quality Control）质量控制， CBR（California bearing ratio）加州承载比

专业英语词汇的数量较多，学生可以通过学习构词法充分挖掘英语词汇之间的内在联系，由少数熟悉的基础英语词汇来速记多数生僻的专业术语，“联系是记忆的桥梁”。利用构词法，结合专业知识，理解、熟悉并记忆专业英语词汇与术语可以达到事半功倍的效果。

1.1.3 语法特点

1. 涉及内容客观

专业英语涉及的内容多为描述客观规律、客观事物和客观现象。这一特点决定了科技人员在撰写科技文献时主要突出表现被描述对象的规律、特性、研究方法、研究成果等，力求客观和准确地陈述，而不需要突出人。因此，专业英语常常使用第三人称语气“It…”结构进行客观的描述。

【例2】 At the next level of traffic generation a single traffic generator could fill a single freeway lane; It is then appropriate to construct a freeway ramp for the exclusive use of the generator without intervening public streets.

译文：在下一个交通生成水平下，由单一的交通发生源产生的交通流可能会充满单条高速公路车道，因此适合设置一条专门为这一发生源所用的匝道而不干扰公共道路。

【例3】 It is easier to make changes in design and to correct errors during construction (and at less expense) if welding is used.

译文：若采用焊接，则在施工阶段更容易修改设计或改正错误（并且花费更少）。

上述两个例句中采用了 It is…的结构，对某些事情或实施过程进行客观的描述，没有加入任何主观色彩。

2. 多用被动结构

由于专业英语的客观性，决定了它非人称的表达方式。读者或许都知道但却不关心动作的执行者是谁。因此，在专业英语中，常使用被动语态。

【例4】 A freeway may be defined as a divided highway facility having two or more lanes for the exclusive use of traffic in each direction and foil control of access and egress.

译文：高速公路可定义为被分隔的两个方向各自具有两条或两条以上汽车专用车道，并且封闭控制进出口的公路设施。

3. 多用非谓语动词结构

非谓语动词包括分词、不定式和动名词，它的使用可以使句子简洁和精练。

【例 5】 Vibrating objects produce sound waves, each vibration producing one sound wave.= Vibrating objects produce sound waves and each vibration produces one sound wave.

译文：振动的物体产生声波，每次振动产生一个声波。

4. 大量使用名词化结构

英语与汉语的一个明显不同点：在汉语中使用动词占优势，动词的使用范围很广；而在英语中使用名词占优势，名词的使用范围很广。英文的科技文体更是如此。其原因是：强调存在的事实，而非某一行为。名词化结构主要是指在专业文献中广泛使用能表示动作和状态的名词，或是起名词作用的非限定动词。

【例 6】 Archimedes first discovered the principle of displacement of water by solid body.
译文：阿基米德最先发现固体排水的原理。

【例 7】 The testing of the air pollution should be considered in highway management.
译文：在公路管理中，应该考虑到进行空气污染测试。

5. 长句较多

道路与桥梁工程研究的目的是揭示交通运输规划、建设、运营及管理的规律，并解释其特点及应用。类似的其他专业工程研究工作是一个复杂的程序，而且程序间的各个环节联系紧密。为了能准确、清晰地表达专业工程中的复杂现象及其之间的紧密关系，其专业文献需要用各种不同的主从复合句，而且会出现复句中从句套从句的现象。

【例 8】 The small tensile strength of concrete as compared with its compressive strength prevents its economical use in structural members that are subjected to tension either entirely (e.g. tie rods) or over part of their cross-section (e.g. beams or other flexural members).

译文：与其抗压强度相比，混凝土的抗拉强度小，妨碍了它在结构构件中的经济实用性，因为结构构件要承受全部拉力（如拉杆）或断面上部的拉力（如梁或其他弯曲构件）。

6. 省略句较多

专业英语为了简洁，有时省略句中的一些成分，如状语从句中的主语和谓语，关联词 which 或 that 引导的定语从句中的关联词和从句中的助动词等。

常见的省略句型如下：

As described above	如前所述
As indicated in Fig.X	如图 X 所示
When necessary	必要时
When needed	需要时

7. 惯用时态

专业英语所描述的内容主要是一般真理或客观规律，所以最常用的时态为一般现在时和现在完成时，除此之外，有时还使用过去时和过去完成时。

【例 9】 This causes concentrated loads that are much larger than the assumed live loads for

which the structure was designed.

译文：这种情况造成的集中载荷比结构设计承受的假定活载荷要大得多。

1.1.4 结构特点

前面的语言、词汇和语法特点属于专业英语领域分析的内容。这些内容形成了专业英语的基础。更进一步，还需要了解专业英语在段落、论文层面上的结构特点，了解隐含在语言运用之中的逻辑思维过程。这样，才有助于把握全篇论文的要点和重点，提高阅读和理解能力。一般在每一自然段落中，总有一个语句概括出该段落的重点。这个语句或在段落之首，或在段落之尾，较少出现在段落中间。若干个自然段会形成一个逻辑（或结构）段落，用以从不同角度来解说某一层面的核心内容。全篇论文则由若干个逻辑段落组成，从不同层面来阐述论文标题所标明的中心思想。实际上，专业文献中通常采用的标题、子标题、编号等形式，就是对论文结构进行逻辑划分。

扫一扫看本节教学课件

1.2 Translation Criteria and Tactics of Professional English 专业英语的翻译标准与策略

专业英语具有其独特的形式及专用语言，一般说来，它有几个基本特点：大量使用术语、公式、数字等，经常借助图表和照片等方式说明其专业内容，结构紧密，主题单一。因此学生在学习专业英语翻译方法时，应该主要从以下几个方面着手：

（1）掌握专业英语的翻译标准；

（2）学会运用适当的翻译技巧，在忠实于原文的基础上，准确地表达其内容；

（3）通过阅读、翻译英文论文摘要等，让学生熟悉专业英语的翻译策略。

1.2.1 专业英语的翻译标准

“信”与“顺”是目前公认的重要的翻译标准。“信”是指准确、忠实，“顺”是指通达顺畅。标准的译文必须在含义上与原文贴切，在行文上读起来流畅。

“信”对专业英语的翻译尤为重要。因为科技图书、报刊和论文的任务在于准确而系统地论述科学技术问题，因而要求有高度的准确性。

【例 10】 Action is equal to reaction, but acts in a contrary direction.

译文一：作用与反作用相等，但它在相反的方向起作用。

译文二：作用与反作用相等，但作用的方向相反。

译文三：作用力与反作用力大小相等，方向相反。

解析：译文一由于拘泥原文结构，语言不够简练通顺；译文二虽然不错，但不如译文三通顺；译文三完全摆脱了原文形式的束缚，并选用四字结构，使译文准确贴切，简洁有力。

【例 11】 In certain cases, friction is an absolute necessity.

原译文：在一定场合下摩擦是一种绝对的必需品。

译文：在某些情况下，摩擦是绝对必要的。

解析：翻译时需要将名词 necessity 转译为形容词“必要的”。可见，有时为了符合汉语行文习惯，需要运用一定的翻译技巧进行适当的变通。

小提示： 从以上这些例句可以看到，“信”与“顺”是辩证统一的。“信”是“顺”的基础，不忠实于原文的译文再通顺也毫无意义；“顺”是“信”的保证，不通顺的译文无疑会影响到译文的质量。因而翻译中必须遵循“信”与“顺”相结合的原则，切实做到使用通顺的语言形式忠实表达原文的思想内容。

1.2.2　专业英语的翻译策略

翻译的过程是读懂并且准确理解原文，然后创造性地使用另一种语言再现原文的过程。对专业英语的翻译要求有高度的准确性，其中的专业术语、公式、图例及数字较多，稍有马虎就会造成南辕北辙的笑话。

翻译的具体策略包括三个层面，即词汇层面、句子层面及篇章段落层面。首先，词汇层面上要求译者能够准确地选择词义、引申词义及掌握词类的转译问题。其次，句子层面上要求译者可以根据英汉两种语言表达方式的不同适当地进行句子成分的改变、合译与分译、增译与省译，掌握被动句、否定句、强调句、从句等特殊句型的翻译方法。最后，篇章段落层面上要求译者在联系上下文对专业英语透彻理解的基础上，根据对原文的理解，使用符合汉语表达的语言形式恰如其分地表达原文内容并进行核对以及对译文语言进一步推敲，即“阅读——理解——表达——校核”的过程。这里着重研究的是篇章段落层面的翻译策略。

1. 阅读

专业英语翻译的突出特点是科学性和专业性，只有对涉及的专业有较深的理解、对该专业的术语、理论和应用有较全面的掌握，才能保证翻译的准确性。

因此，首先要了解专业内容，掌握一定的专业英语词汇量，按照专业逻辑大致读懂原文。

2. 理解

理解主要是通过联系上下文进行的，对英文的透彻理解是翻译的基础和关键。通常应注意两个方面：一是正确地理解原文的词汇含义、句法结构和习惯用法；二是要准确地理解原文的逻辑关系。

【例 12】 Pavement are classified as “rigid” or “flexible”, depending on how they distribute surface loads.

原译文： 路面被分为“坚硬的”或“柔韧的”，这要取决于它们怎样传递表面载荷。

译文： 路面按照它们传递表面载荷情况可分为“刚性的”或“柔性的”。

解析： rigid 译为“坚硬的”，意思上不错，但不符合专业术语行文的习惯，应改为“刚性的”更妥当。由此可见，在选择词义时，必须从上下文联系中去理解词义，从专业内容上去判断词义。

3. 表达

表达就是译者根据对原文的理解，使用汉语恰如其分地表达原文的内容。在表达阶段最重要的是表达手段的选择，同一个句子的翻译可能有好几种不同的译法，但在质量上往往有高低之分。

【例 13】 Sometimes entry and exit roads are curved in a regular pattern of circles to and from each level.

原译文：有时立体交叉的入口路段和出口路段被弯曲成匀称的圆形与各层立交路相接。

译文：有时立体交叉的入口路段和出口路段与各层立交路以匀称的圆弧形路径引入、引出。

解析：原译文为直译，语言不够通顺；改后的译文流畅贴切。

4. 校核

校核是理解和表达的进一步深化，是对原文内容进行核对以及对译文语言进一步推敲。

【例 14】 One of the common defects in bridge maintenance is the periodic addition of surface dressing resulting in dead loads much in excess of original design.

原译文：桥梁养护中普遍存在的问题之一是路面不断增加，导致静载荷大大超过原设计。

译文：……路面因整修不断增厚……

1.3 Guide to Writing Scientific Abstract 科技论文摘要写作指南

1.3.1 科技论文的体例

国际标准化组织（International Organization for Standardization）、美国国家标准化协会（American National Standard Institute）和英国标准协会（British Standards Institute）等国际组织都对科技论文的写作体例（stylistic rules）作出了规定，其基本内容介绍如下。

对期刊类科技论文，主要部分包括：

（1）Title 题目；

（2）Authorship 作者信息；

（3）Abstract 摘要；

（4）Keywords 关键词；

（5）Introduction 引言；

（6）Method, material, equipment and test (experiment) procedure 介绍试验方法、材料、设备和试验步骤等；

（7）Results and Discussions 结果和讨论；

（8）Conclusions 结论；

（9）Acknowledgements 致谢；

（10）References 参考文献；

（11）Appendix 附录。

以上只是对期刊类科技论文做出的框架规定，在实际写作过程中允许根据实际情况适当调整。

1.3.2 标题与署名

论文标题属于特殊文体，一般不采用句子，而是采用名词、名词词组或名词短语的形

式，通常省略冠词。

具体要注意以下几点：

（1）准确、简明地表现论文的主题和内涵；

（2）尽量使用名词化短语，10～20 个字，字数控制在两行；

（3）单词的选择要规范化，要便于文献检索，如利用题录、索引、关键词等二次文献信息查找文献线索和根据文献线索查找原始文献。

【例 15】 Bridge Live-load Model 桥梁活载荷模型

【例 16】 Current Durability Situation of Concrete Highway Bridge Structures in Southeast Coastal Areas 东南沿海地区混凝土公路桥梁结构的耐久性现状

1.3.3　摘要

摘要是论文的核心体现，直接影响读者对论文的第一印象。

1. 摘要的基本特点

（1）能使读者理解原文的基本要素，如研究目的、方法、结果、结构和意义，能脱离原文而独立存在。

（2）是对原文的精华提炼和高度概括，信息量较大。

（3）具有客观性、准确性和文献检索功能。

2. 形式和内容

（1）若无特殊规定，摘要位于论文标题和正文之间，偶尔也出现在正文之后。

（2）篇幅控制在 150～200 个英文词。对长篇报告或学位论文，一般在 250 个词左右，最多时不超过 500 个词。

（3）通常不分段。对长篇报告或学位论文可以分段，但段落不宜多。

（4）与标题写作相反，要用完整的句子，不能使用短语。另外，要注意使用连词连接前后句，避免行文干涩单调。

（5）避免使用暂时还不熟悉的或容易引起误解的单词缩写和字符；不可避免时要在第一次出现处加以说明，如 CCES（Chinese Civil Engineering Society）。

（6）少用或不用第一人称，多用第三人称被动语态，以体现客观性。

（7）避免隐晦和模糊的表达，采用准确、简洁的语句概括全文的目的、意义、观点、方法和结论。

（8）注意摘要的独立性与完整性，摘要的内容与结论必须与原文一致。

（9）通常摘要采用一个主题句（Topic sentence）开头，以阐明论文主旨，或引出论文的研究对象，或铺垫论文的工作，避免主题句与论文标题的完全或基本重复。

（10）在摘要后，通常要附上若干个表达全文内容的关键词、主题词或检索词（indexing term），应使用规范化、普遍认可的单词和术语。

3. 摘要常用句型

在撰写摘要时，可套用一些固定的句型。掌握摘要常用句型的特点，并结合实际情况灵活运用，这一点非常重要。

(1) This paper (the author(s)): introduce(s), propose(s), present(s), describe(s), discuss(es), deal(s) with, bear (s) on, show (s) …

(2) In this paper …is (are) introduced/ proposed/ presented/ described/ discussed/ studied …

(3) This paper is mainly concerned with / aimed (mainly) at/ intended to + the study/ determination of/ computation …

(4) The chief (aim/ main purpose/ primary object/ primary objective) of (the present study is / this investigation was/ our research has been/ these studies will be) to (obtain some results/ review the process/ access the role/ find out what/ reveal the cause of/ establish the equation) …

(5) …has (have) (been) concluded/gained/ obtained/ yielded/ arrived at/ generated/ acquired/ achieved …

4. 摘要示例

Abstract: The investigation of concrete highway bridge structures in Zhejiang Province is carried out for an overall knowledge of current durability situation in southeast coastal area. Several representative bridges are selected for detailed inspection. The inspection and investigation results indicate that durability degradation varies in degree and style with environmental conditions of different areas significantly. Consequently a durability environmental zonation standard is required for concrete structural design. Some preliminary suggestion on durability design, construction and evaluation of coastal highway bridge structures are also proposed based on the analysis.

Keywords: Highway bridge; durability; environmental zonation; concrete structure.

摘要：为全面了解东南沿海地区混凝土公路桥梁结构的耐久性现状，在浙江省内开展了混凝土公路桥梁结构的耐久性现状调查，并对典型桥梁进行了详细检测。调查和检测的结果表明，这些结构普遍存在耐久性老化的现象，并且随环境条件的不同在老化程度与老化类型上表现出明显的地区差异。因此建立混凝土结构耐久性环境区划标准对结构耐久性设计十分必要。在分析基础上对东南沿海地区混凝土公路桥梁结构耐久性设计、施工与评估提出了若干具体初步建议。

关键词：公路桥梁；耐久性；环境区划；混凝土结构

知识分布网络

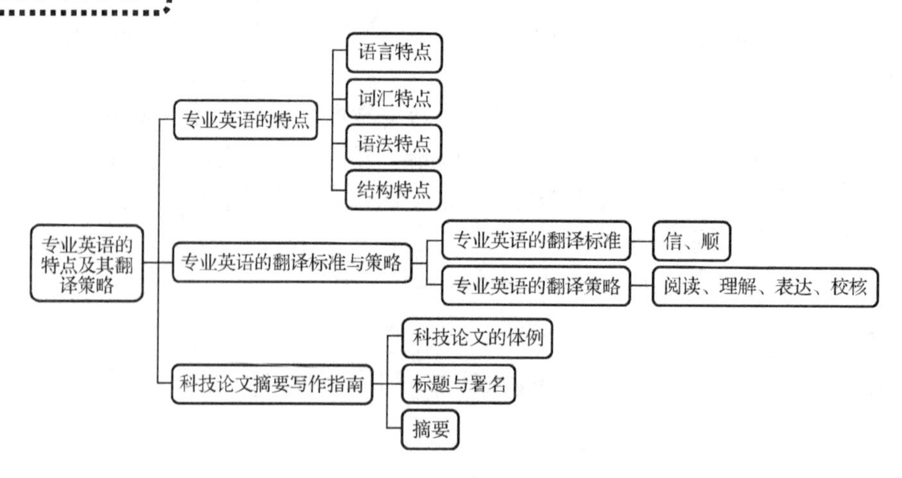

知识梳理与总结

通过对本章的学习，能够系统地掌握市政工程等专业英语的文体特征和专业文献的阅读方法，掌握英语资料阅读、翻译以及英文摘要写作的技巧。本章必须掌握的知识点有：

（1）专业英语的基本特点；

（2）学会运用适当的翻译方法和技巧，在忠实于原文的基础上，准确地表达其内容；

（3）英文论文摘要写作及阅读的方法。

希望学生结合本章内容，通过参考资料与观察总结，理解市政工程等专业词汇和专业术语的学习方法，自我培养专业英语的翻译能力。

思考与练习题 1

扫一扫看第 1 章习题答案

Task 1: 阅读市政工程等专业的英文文献（如学术期刊论文），体会专业英语的基本特点。

Task 2: 阅读下列段落，分析其中的语法和词汇特点，并尝试翻译成中文。

(1) Digital Image Analysis (DIA) has been widely studied as a means of automating aggregate tests. In DIA, an aggregate sample from the production stream is photographed with a camera; this image is then digitized for computer analysis. To extract size information on each particle in the digital image, algorithms for image segmentation and size measurement are used. That is, after the particles in the image are separated by the segmentation algorithm, all of the particles are measured, one by one, in a computationally intensive manner.

(2) Taking a reinforced concrete simply supported T-shaped beam bridge in Daqing city as the research object, to analyze the existing disease and put forward reasonable reinforcement measures. Then, to establish the space finite element model of simply supported T beam bridge by using the finite element software Midas Civil and carry out the bearing capacity checking of the reinforcement bridge structure. The calculation results show that the reinforcement scheme has certain feasibility.

Task 3: Translate the following sentences into Chinese.

(1) A prediction of the duration of the period when building materials cannot be supplied would be of value in the planning of construction.

(2) Force is any push or pull that tends to produce or prevent motion.

(3) The practice of employing rubber reinforced as bridge supports to allow for temperature movements and variations in foundation is well established, and about half of all new bridges are now built in the U. K. in this way.

(4) Grouting of the tendons usually follows the freedom of the ducts from obstructions.

(5) The broken bridge is now back to traffic.

Task 4: Translate the following abstract into English.

摘要：装配式预应力混凝土空心板桥因其具有诸多优点而在我国桥梁建设中被广泛应用，然而传统的横向铰接空心板桥在实际运营中常出现铰缝被损坏的现象，导致空心板桥

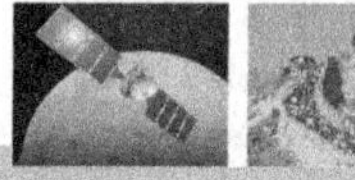

在运营时的受力和设计时的受力不一致。在总结了国际上常见的几种装配式空心板横向连接形式的基础上提出一种采用横隔板作为主要连接构件的横向连接方案，使装配式空心板桥的横向传力更加明确。分析计算表明，横隔板的数量对空心板的横向分布和横向变形影响很小，采用在横隔板中施加预应力的方式可以从结构上保证横隔板受力的可靠性。

关键词：装配式空心板，铰缝损坏，横隔板，横向连接，可靠性

Chapter 2 General Knowledge on Civil Engineering

第 2 章 土木工程专业基础

教学导航

教	知识重点	1. Construction Materials 建筑材料; 2. Highway Surveying Technology 道路测量技术; 3. Structural Analysis 结构设计; 4. Tender Document and Contracts 标书与合同
	知识难点	英文标书与合同的撰写
	推荐教学方式	明确教学目标，依据教材、多媒体课件、视频等教学手段，让学生掌握建筑材料、测量学、力学、标书与合同等土木工程专业基础的英语知识
	建议学时	10 学时
学	推荐学习方法	以任务驱动和小组讨论的学习方式为主，课堂学习与自主学习相结合
	必须掌握的理论知识	建筑材料、测量学、力学、标书与合同等相关专业英语知识与词汇
	必须掌握的技能	通过对本章的学习，能够阅读与土木工程专业基础相关的英文文献、学术论文等，尝试撰写英文标书与合同

2.1 Construction Materials 建筑材料

扫一扫看本节参考译文

2.1.1 Text

扫一扫看本节教学课件

In previous years, the principal construction materials were wood and masonry-brick, stone, tile and other materials. The courses or layers were bound together with mortar or bitumen or some other binding agent. The Greeks and Romans sometimes used iron sticks to strengthen their building. The Romans also used a natural cement called pozzolana (volcanic ash), which became as hard as stone under water.

Construction aggregate

Construction aggregate (or simply "aggregate") is a broad term of coarse to medium grained particulate material used in construction, which is usually hard, inert material such as sand, gravel, crushed stone, slag, rock dust and recycled concrete. Aggregate is a component of composite material such as concrete and asphalt concrete; the aggregate serves as reinforcement to add strength to the overall composite material (Fig.2-1). Properly selected and graded aggregates are mixed with the asphalt to form the *HMA*[1] pavements. Aggregates are the principal load-supporting components of the HMA pavements, totaling approximately 95 percent of the mixture by weight.

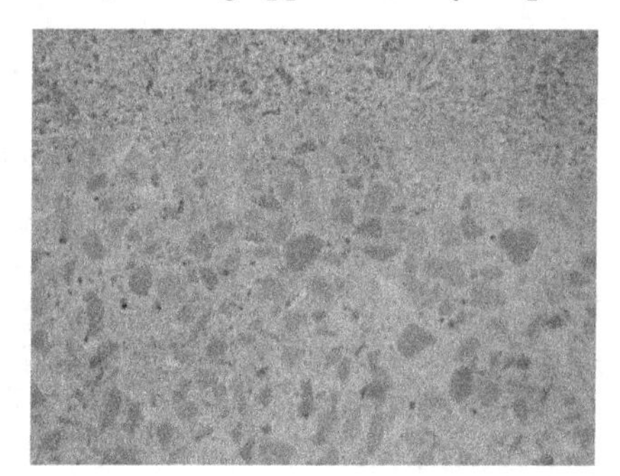

Fig.2-1 Grinding concrete exposes aggregate stones.

Sources of construction aggregate can be classified into three main areas:

1. Natural aggregate

Natural aggregate can be obtained from mining of mineral aggregate deposits, including sand, gravel, and stone (Fig.2-2, Fig.2-3).

2. Processed aggregate

When natural aggregate has been crushed and screened to make it suitable for HMA pavements, it is considered a processed aggregate. Crushing typically improves the particle shape by making rounded particles more angular.

For example, crushed stone is a processed aggregate. It is created when the fragments of large stones are crushed so that all particle faces are fractured. Variation in size of particles is achieved by screening (Fig.2-4, Fig.2-5).

Fig.2-2　Limestone quarry.

Fig.2-3　A gravel and sand extraction facility

Fig.2-4　10 mm graded aggregate

Fig.2-5　20 mm graded aggregate

3. Synthetic aggregate

Aggregate produced by altering both physical and chemical properties of a parent material is called synthetic or artificial aggregate, which is the byproduct of manufacturing and a final burning process, such as the production of iron or the burning of domestic refuse. Blast furnace slag is an example of a synthetic aggregate.

Asphalt

The black binding agent known as asphalt has been used for road construction for centuries. Deposits of asphalt and rock asphalt exist naturally, but the most used today is produced by means of refining crude oil. Asphalt is a constituent of petroleum and is isolated through the refining process (Fig.2-6).

Asphalt is called a bituminous material because it contains *bitumen*[2], a hydrocarbon material soluble in carbon disulfate. The tar obtained from the destructive distillation of soft coal also contains bitumen. Both petroleum asphalt and coal tar are called bituminous materials. Because of great difference in their properties, petroleum asphalt should not be confused with coal tar. Petroleum asphalt is composed almost entirely of bitumen while the bitumen content in coal tar is relatively low.

Fig.2-6 Refined asphalt

One of the characteristics and advantages of asphalt as engineering construction and maintenance material is its versatility. Although semi-solid at ordinary temperature, asphalt may be liquefied by heat, dissolving it in solvents, or emulsifying it. Asphalt is a strong cement that is readily adhesive and highly waterproof and durable, making it particularly useful in road construction. It is also highly resistant to the actions of most acids, alkalis, and salts.

The semi-solid form known as asphalt binder is broadly used in Hot-Mix Asphalt (HMA) pavements. Asphalt binder is produced in various types and grades, such as emulsified asphalt, cut-back asphalt, sulphur extended asphalt and so on.

Steel

Steel is an alloy of iron, a small amount of carbon and other elements. Because of its high tensile strength and low cost, it is a major component used in buildings, infrastructure, highways, bridges (Fig.2-7), tools, machines and weapons. Most large modern structures, such as stadiums and skyscrapers, bridges, and airports, are supported by a steel skeleton. Some concrete structures employ steel for reinforcing. The advantage of steel is its tensile strength—it does not lose its strength when it is under a calculated degree of tension which tends to pull apart many materials. New alloys have further increased the strength of steel and eliminated some problems, such as fatigue, which is a tendency to weaken structure as a result of continual changes in stress.

Fig.2-7 Steel bridge panorama

Portland cement and concrete

Modern cement, called Portland cement, was invented in 1824. Portland cement is by far the most common type of cement in general use around the world. Limestone and clay is milled to powder and mixed in the appropriate proportion. Cement is produced by heating the mixture of limestone (calciferous material) and clay (silica and alumina) to 1 450 ℃ in a kiln, whose process is known as calcination. Then the resulting sinter is cooled and ground to form coarse clinker, which is the basic ingredient of ordinary Portland cement (OPC). This material is a mixture of calcium aluminate and calcium silicate, both of which react with water to form the stone-like mass. Portland cement may be grey or white.

A characteristic of cement which is of great concern in highway engineering is the shrinkage which occurs in the hardening material, during and after curing. As curing proceeds, some water is absorbed within the cured material, causing the cement particles to swell.

Portland cement is a basic ingredient of concrete, mortar and most grout. The most common use for Portland cement is in the production of concrete. Concrete is a composite material consisting of aggregate (small stone, crushed rock, gravel and sand), cement and water. Different proportions of the ingredients produce concrete with different strength and weight. Concrete is very versatile; it can be poured, pumped, or even sprayed into all kinds of shapes. As a construction material, concrete can be cast in almost any shape desired, and once hardened, can become a structural (load bearing) element.

Reinforced concrete

Reinforced concrete (RC) is a composite material in which relatively low tensile strength and ductility of concrete are counteracted by the inclusion of reinforcement with higher tensile strength and ductility. The reinforcement is usually reinforcing steel bars (rebars) and is usually embedded passively in the concrete before the concrete sets (Fig.2-8). Reinforcing schemes are generally designed to resist tensile stresses in particular regions of the concrete that might cause unacceptable cracking and structural failure. Modern reinforced concrete can contain varied reinforcing materials made of steel, polymers or alternate composite material in conjunction with rebars or not. Reinforced concrete may also be permanently stressed (in tension), so as to improve the behavior of the final structure under working loads. In the United States, the most common methods of doing this are known as *pre-tensioning*[3] and post-tensioning.

2.1.2 Notes

1. Hot-Mix Asphalt concrete (commonly abbreviated as HMA): This is produced by heating the asphalt binder to decrease its viscosity, and drying the aggregate to remove moisture from it prior to mixing. Mixing is generally performed with the aggregate at about 300 ℉ (roughly 150 ℃) for virgin asphalt and 330 ℉ (166 ℃) for polymer modified asphalt, and for the asphalt cement at 200 ℉ (93 ℃). Paving and compaction must be performed while the asphalt is sufficiently hot. In many countries paving is restricted to summer months because in winter the compacted base will

Fig.2-8 A heavy reinforced concrete column, seen before and after the concrete has been cast in place around the rebar cage.

cool the asphalt too fast before it is able to be packed to the required density. HMA is the form of asphalt concrete most commonly used on high traffic pavements such as those on major highways, racetracks and airfields. It is also used as an environmental liner for landfills, reservoirs, and fish hatchery ponds.热拌沥青混合料

2. 单词区分 asphalt/bitumen:

Asphalt is a black substance used to make the surfaces of things such as roads and playgrounds.沥青；柏油

Bitumen is a black sticky substance which is obtained from tar or petrol and is used in making roads.沥青；柏油

3. Pre-tensioning is usually with sophisticated behaviors and small stiffness.

用预张法在张拉过程中的受力复杂，刚度较小。

2.1.3 New Words and Expressions

masonry [ˈmeɪsənri]	（筑墙或盖楼用的）砖石
mortar [ˈmɔ:tə(r)]	灰浆；砂浆；胶泥
bitumen [ˈbɪtʃəmən]	沥青；柏油
binding agent	黏合剂，结合剂
cement [sɪˈment]	水泥，胶合剂
pozzolana [ˌpɒtswə'lɑ:nɑ:]	火山灰（可用作水泥原料）
volcanic ash	火山灰
aggregate	骨料；集料（可用于混凝土或修路等）；混凝料
inert	【化】惰性的
crushed stone	碎石
slag	矿渣，熔渣

rock dust	岩粉
concrete	混凝土
asphalt concrete	沥青混凝土
reinforcement	加固
graded aggregate	级配骨料，分级粒料
asphalt [ˈæsfælt]	沥青；柏油
natural aggregate	天然骨料，天然集料
limestone quarry	石灰石采矿场
processed aggregate	加工骨料，加工集料
screen	筛分，筛选
fracture[ˈfræktʃə]	（使）折断，破碎
synthetic aggregate	合成骨料
artificial aggregate	人工合成骨料
parent material	原材料，母料
byproduct	副产品
domestic refuse	生活垃圾
blast furnace	鼓风炉
rock asphalt	岩沥青
refine [rɪˈfaɪn]	提炼；精炼
crude oil [kru:d ɔil]	原油
petroleum [pəˈtrəʊliəm]	石油
bituminous [bɪˈtju:mɪnəs]	含沥青的
hydrocarbon [ˌhaɪdrəˈkɑ:bən]	【化】碳氢化合物，烃
soluble [ˈsɒljəbl]	【化】可溶的
carbon disulfate	二硫化碳
tar [tɑ:]	焦油，沥青，柏油
destructive distillation	分解蒸馏，干馏
soft coal	烟煤
petroleum asphalt	石油（地）沥青
coal tar	煤焦油
semi-solid	半固体，半固态的
at ordinary temperature	常温
liquefy [ˈlɪkwɪfaɪ]	（使）液化，溶解
dissolve [dɪˈzɒlv]	（使）溶解
solvent [ˈsɒlvənt]	【化】溶剂
emulsify [ɪˈmʌlsɪfaɪ]	使乳化
waterproof [ˈwɔ:təpru:f]	adj.不透水的，防水的；vt.使防水，使不透水

road construction	筑路，道路建筑（施工，工程）
acid [ˈæsɪd]	【化】酸；酸性物质
alkali [ˈælkəlaɪ]	【化】碱
salt	【化】盐
asphalt binder	沥青结合料
emulsified asphalt	乳化沥青
cut-back asphalt	【化】稀释沥青；油溶沥青；轻制沥青
sulphur extended asphalt	掺硫沥青
alloy [ˈælɔɪ]	合金
carbon [ˈkɑ:bən]	【化】碳
tensile strength	抗张强度
infrastructure	基础设施；基础建设
highway	（尤指城镇间的）公路，干道，交通要道
stadium [ˈsteɪdiəm]	运动场；体育场
skyscraper [ˈskaɪˌskrepɚ]	摩天大楼，超高层大楼
skeleton [ˈskelɪtn]	（建筑物等的）骨架
fatigue [fəˈti:g]	疲劳
panorama [ˌpænəˈrɑ:mə]	全景画；全景照片
Portland cement	【交】硅酸盐水泥
ordinary Portland cement (OPC)	普通硅酸盐水泥
calcium aluminate	铝酸钙
calcium silicate	硅酸钙
react	（使发生）相互作用；（使起）化学反应
highway engineering	公路工程（学）
shrinkage [ˈʃrɪŋkɪdʒ]	收缩
cured material	固化材料
swell [swel]	膨胀
grout [graʊt]	薄泥浆，水泥浆
composite material	合成材料，复合材料
pour	倾泻，倾倒
pump	泵；用泵输送
spray	喷洒，喷射
cast	铸型，铸造
load bearing	承载，承重
reinforced concrete (RC)	钢筋混凝土
ductility [dʌk'tɪlɪtɪ]	延展性，韧性
reinforcement	加固，加强

reinforcing steel bar	钢筋
rebar	钢筋，螺纹钢筋
embed[ɪm'bed]	把……嵌入
crack [kræk]	断裂，开裂，裂缝
polymer ['pɒlɪmə(r)]	多聚物;【高分子】聚合物
tension	【物】张力，拉力
pre-tensioning	预张法
post-tension	后拉，后张
cast in place	浇筑，现场浇筑
the rebar cage	钢筋笼

扫一扫看本节参考译文

2.2 Highway Surveying Technology 道路测量技术

扫一扫看本节教学课件

2.2.1 Text

Highway surveying techniques have been revolutionized in the past decade because of the rapid development of electronic equipment and computers. Techniques for highway surveying can be grouped into three general categories: conventional ground surveys, remote sensing, and computer graphics.

Ground Surveys

Conventional ground surveys are the basic location techniques for highway engineers until developments in electronics. The most important equipments applied for ground surveys are the <u>theodolite</u> for measuring angles in both vertical and horizontal planes, the <u>level</u> for measuring height differences, and the <u>tape</u> for measuring horizontal distances. However, *total station*[1] is now used for most surveys.

Total station or TST (Total Station Theodolite) is an electronic/optical instrument used for surveying (Fig.2-9). The total station is an electronic theodolite (transit) integrated with an electronic distance measurement (EDM) <u>to read angles</u> at both vertical and horizontal planes, to distances and height differences from the instrument to a particular point, <u>to collect data</u> by internal electronic data storage and <u>to determine the coordinates</u> (*X*, *Y*, and *Z* or easting, northing and elevation) of an unknown point relative to a known coordinate as long as a direct line of sight can be established between the two points.

The standard **theodolite** consists of a <u>telescope</u> with vertical and horizontal cross hairs, a <u>graduated arc or vernier</u> for reading vertical angles, and a <u>graduated circular plate</u> for reading horizontal angles, whereas the electronic theodolite provides a digital readout of those angles. These readouts are continuous and angles can be checked at any time. The telescope on both instruments is set so that it can rotate vertically round a horizontal axis. Of the standard theodolite, two vertical arms support the telescope on its horizontal axis, with the graduated arc attached to one

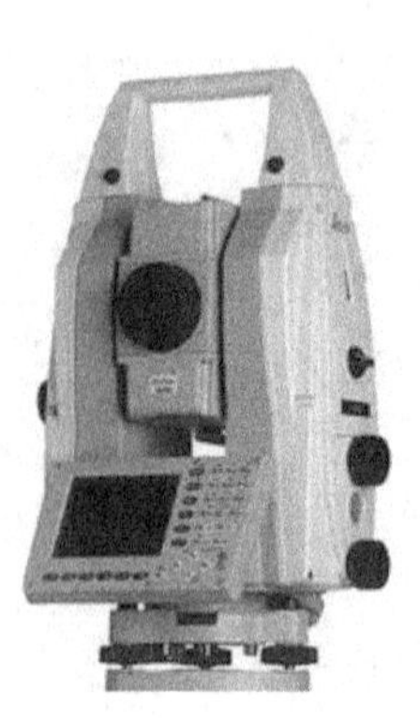

Fig.2-9 Total station

of the arms. The arms are attracted to a circular plate which can rotate horizontally with reference to the graduated circular plate, thereby providing a means for measuring horizontal angles (Fig.2-10).

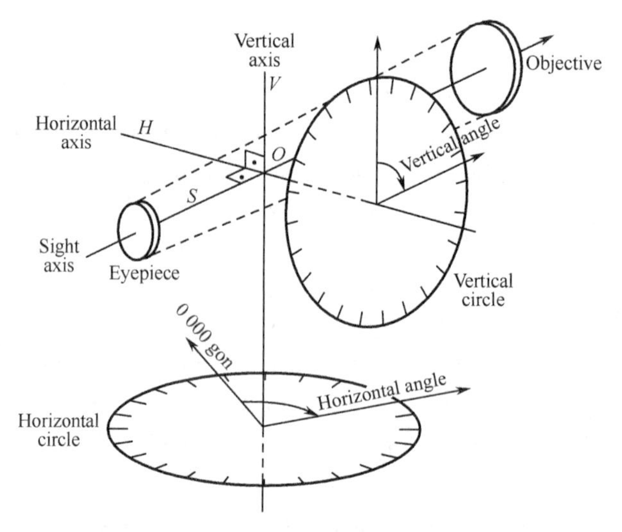

Fig.2-10 The axes and circles of a theodolite

An **electronic distance measurement (EDM)** device consists mainly of a transmitter located at one end of the distance to be measured and a reflector at the other end. The transmitter sends a light beam, a low-power laser beam or a high-frequency radio beam in the form of short waves which are reflected back to the transmitter. The difference in phase between the transmitted and reflected beams is measured electronically and used to determine the distance between the transmitter and the reflector. This equipment can measure distances up to about 1 000 meters in atmospheric conditions.

The essential parts of a *level*[2] are the telescope with vertical and horizontal cross hairs, a level bar, a spindle, and leveling head. The level bar on which the telescope is mounted is rigidly fixed to the spindle. The level tube is fixed to the telescope or the level bar so that it is parallel to the telescope. The spindle is fitted into the leveling head in such a way that allows the level to rotate round the spindle as an axis, with the leveling head attached to a tripod. The level also carries a bubble that indicates whether the level is properly centered. The centering of the bubble is done by using the leveling screws provided. Digital levels are increasingly common in replacing conventional spirit levels particularly in civil engineering applications, such as highway construction and steel structure erection, for on-site angle alignment and leveling tasks (Fig.2-11).

Fig.2-11　Automatic level in use on a construction site

Remote Sensing

Remote sensing[3] is the measurement of distances and elevations by means of devices located above the earth, such as airplanes or orbiting satellites using Global Positioning Satellite systems (GPS). The most usually used remote-sensing method is photogrammetry which utilizes aerial photography (Fig.2-12). Photogrammetry is the science of obtaining accurate and reliable information through measurements and interpretation of photographs, displaying this information in digital form and map form. This process is fast and economical for large projects but can be very expensive for small projects. The break-even area for which photogrammetry can be used varies from 120 to 400 square kilo meters. The available use of the method depends on the type of terrain. Difficulties will appear when it is used for terrain with the following characteristics.

Fig.2-12　An aerial survey camera externally mounted on a Cessna 172 capturing photographs used in digital mapping

- Thick forest areas, such as tropical rain forests, completely covering the ground surface.

- Areas with deep canyons or tall buildings, possibly concealing the ground surface on the photographs.
- Areas with uniform shades, such as plains and some deserts.

The most common usages of photogrammetry in highway engineering are the determination of suitable locations for highways and the preparation of base maps for design mapping, showing all natural and man-made features and contours with 2 or 5 feet intervals. Then, the first task is to obtain the aerial photographs of the area if none is available. The information on the aerial photography is then used to convert these photographs into maps.

Computer Graphics

Computer graphics, when applied for highway location, is commonly the combination of photogrammetry and computer techniques. The information obtained from photogrammetry is stored in a computer, which is linked with a stereo-plotter and computer monitor. With input of the appropriate command, the terrain model or the horizontal and vertical alignment are obtained and displayed on the monitor. It is therefore easy to change some control points and obtain a new alignment of the highway on the screen, permitting the designer to immediately see the effects of any changes.

2.2.2 Notes

1. 全站仪，即全站型电子测距仪（total station），是一种集光、机、电于一体的高性能测量仪器，是集水平角、垂直角、距离（斜距/平距）、高差测量功能于一体的测绘仪器。

2. 水准仪（level）是建立水平视线测定地面两点间高差的仪器。原理为根据水准测量原理测量地面两点间的高差。主要部件有望远镜、管水准器（或补偿器）、垂直轴、基座、脚螺旋。按结构分为微倾水准仪、自动安平水准仪、激光水准仪和数字水准仪（又称电子水准仪）；按精度分为精密水准仪和普通水准仪。

3. 遥感技术是 20 世纪 60 年代兴起的一种探测技术，是根据电磁波的理论，应用各种传感仪器对远距离目标所辐射和反射的电磁波信息，进行收集、处理、最后成像，从而对地面各种景物进行探测和识别的一种技术。把遥感器放在高空气球、飞机等飞行器上进行遥感，称为航空遥感。把遥感器装在航天卫星上进行遥感，称为航天遥感。航空和航天遥感技术广泛应用于国民经济、军事等很多方面，例如气象观测、资源考察、地图测绘和军事侦察等。

2.2.3 New Words and Expressions

survey [ˈsɜːveɪ]	测量；勘测；测绘
electronic equipment	电子设备
ground survey	地面测量
remote sensing	遥感
computer graphics [ˈɡræfɪks]	计算机图形学
location technique	定位技术，the act of finding the position

electronics [ɪˌlekˈtrɒnɪks]	电子学
theodolite [θiˈɒdəlaɪt]	经纬仪
vertical [ˈvɜ:tɪkl]	adj.垂直的；n.垂直线，垂直面
horizontal [ˌhɒrɪˈzɒntl]	adj.水平的；地平线的；n.水平线；水平面
plane	【几何】平面，flat or level surface
level	水平仪，水准仪
tape	卷尺
total station	全站仪
total station theodolite	全站仪
electronic [ɪˌlekˈtrɒnɪk]	电子的
optical [ˈɒptɪkl]	视觉的，光学的
transit [ˈtrænzɪt]	经纬仪，theodolite
electronic distance measurement	电子测距仪
coordinate [kəʊ'ɔ:dɪneɪt]	【数】坐标
elevation [ˌelɪˈveɪʃn]	高度，海拔
telescope [ˈtelɪskəʊp]	望远镜
cross hair	交叉瞄准线，【光】标线，十字线
graduated arc [ˈgrædjʊeɪtɪd ɑ:k]	分度弧
vernier ['vɜ:nɪə]	游尺，游标，游标尺
graduated circular	分度圆
horizontal/vertical angle	水平角/垂直角
digital [ˈdɪdʒɪtl]	数字的
readout	读出器，读出
horizontal axis	水平轴
transmitter	发射器
reflector	反射器
light beam	光束
low-power laser beam	低能量激光束
phase [feɪz]	【物理学】相位
slope [sləʊp]	斜坡；斜面；倾斜；斜率
level bar	水平杆，水平尺
spindle [ˈspɪndl]	轴
leveling head	校平头
level tube	水准测管，水准器，水准仪管
parallel [ˈpærəlel]	平行的
tripod [ˈtraɪpɒd]	【摄】三脚架
bubble [ˈbʌbl]	泡，水泡

leveling screw	校平【水准】螺旋；准平螺钉
digital level	数字水准仪
spirit level	（气泡）水准仪
on-site angle alignment	现场角度核准
remote sensing	遥感技术
airplane	飞行器，飞机
orbiting satellite	轨道运行卫星
Global Positioning Satellite systems	GPS，全球定位卫星系统
photogrammetry[fəʊtə'græmətrɪ]	摄影测量法；摄影测量学
aerial photography	航空摄影
break-even [breik ˈi:vən]	收支平衡；不赔不赚；经济的
terrain [təˈreɪn]	地形，地势；地面，地带；【地理】岩层
tropical rain forest	热带雨林
canyon [ˈkænjən]	峡谷
contour [ˈkɒntʊə(r)]	（地图上表示相同海拔点的）等高线
interval [ˈɪntəvl]	间隔；【数学】区间
aerial survey camera	航空测量摄影仪
mount	安装
Cessna	cessna aircraft company，塞斯纳飞机公司
digital mapping	数字制图
highway location	公路定线
computer technique	计算机技术
stereo-plotter	立体绘图仪
computer monitor	计算机显示器
terrain model	地面模型
horizontal alignment	水平定线，水平线型
vertical alignment	竖向定线，竖向线型
control point	控制点，检测点

2.3 Structural Design 结构设计

扫一扫看本节参考译文

2.3.1 Text

扫一扫看本节教学课件

An engineering project basically has three stages: planning, design and construction.

Structural design involves determining the most suitable proportions of a structure and dimensioning the structural elements. This is a stage in need of advanced techniques and mathematical modeling during a structural engineering project. The successful designer is at all

times fully conscious of the various factors that are involved in the preliminary planning for the structure, likewise, of the various problems that may be encountered later in the construction.

Loads may be broadly classified as permanent loads that are constant in magnitude and remain in one position and variable loads that may change in magnitude and position. Permanent loads are also referred to as dead loads which include the self weight of the structure and other loads permanently attached to the structure (Fig.2-13). Variable loads are also referred to as live (or imposed) loads (Fig.2-14), and include those caused by construction operations, wind, earthquakes, snow, blasts, and temperature changes in addition to those that are movable (i.e. vehicles, trains, etc.).

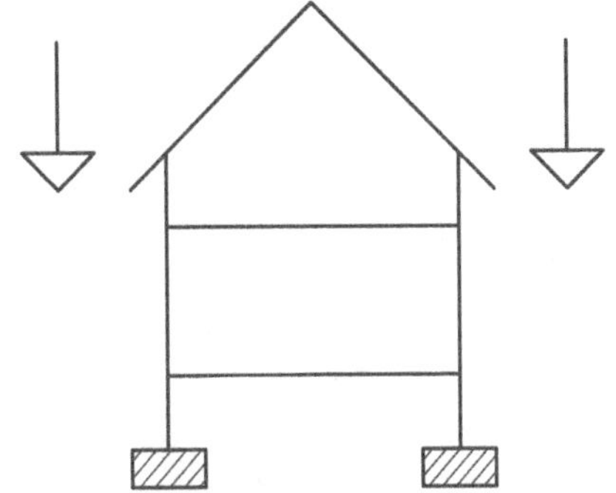

Fig.2-13　Permanent loads/dead loads

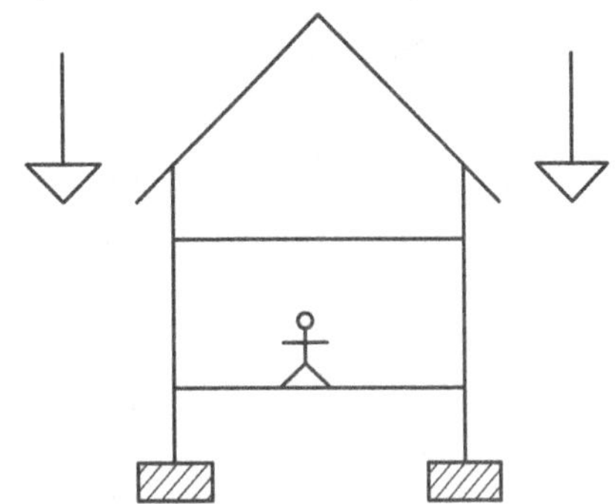

Fig.2-14　Imposed loads

Wind loads act as pressures on windward surfaces and pressures or suctions on leeward surfaces. Impact loads are caused by suddenly applied loads or by the vibration of moving or movable loads. Earthquake loads are forces caused by the acceleration of the ground surface during an earthquake.

The increased load would (1) cause a fatigue or a buckling or a brittle-fracture failure or (2) just produce yielding at one internal section or (3) cause elastic-plastic displacement of the structure or (4) cause the entire structure to be the point of collapse (Fig.2-15).

Fig.2-15　Broken Tacoma Narrows Bridge by wind loads

A structure that is initially static and remains static when acted upon by applied loads is said to be in a state of equilibrium. The resultant of the external loads on the body and the supporting forces or reactions is zero.

Loads cause stresses, deformations, and displacements in structures. Assessment of their

effects is carried out by the method of structural analysis. Excess load or overloading may cause structural failure, and hence such possibility should be both considered in the design and strictly controlled. First, the design of any structure involves determining loads and other design conditions must be resisted by the structure and therefore must be considered in its design. Then comes the analysis of the internal gross forces (thrust, shears, bending moments, and twisting moments) , stress intensities(Fig.2-16, Fig.2-17), strains, deflections, and reactions produced by the loads, temperature, *shrinkage, creep*[1], or other design conditions. Finally comes the proportioning and selection of materials of the members and connections so as to resist adequately the effects produced by the design conditions.

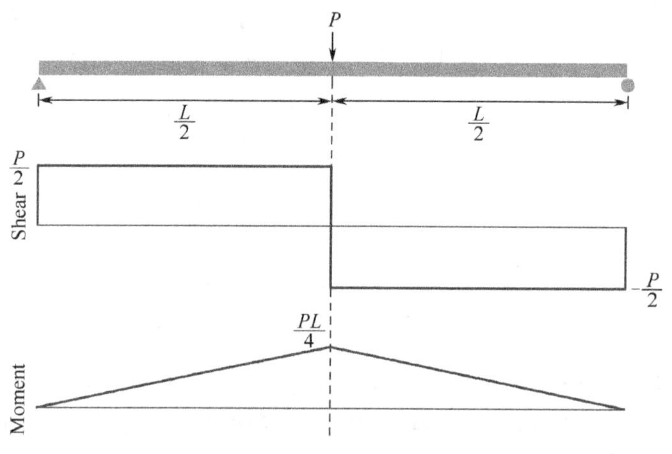

Fig.2-16 Shear and moment diagram for a simply supported beam with a concentrated load at mid-span.

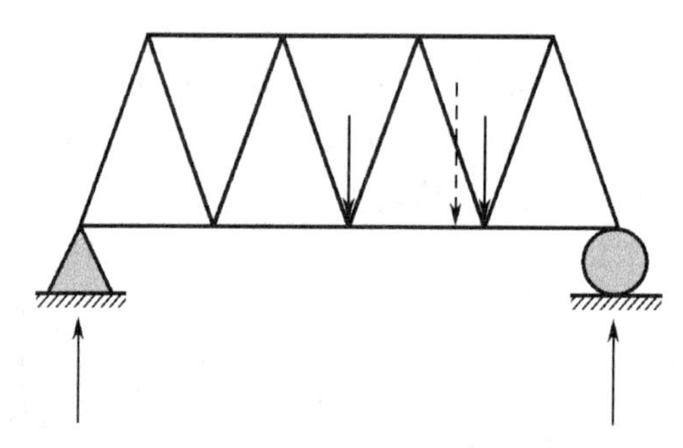

Fig.2-17 Truss modeling under uniform stress

For civil engineering structures such as bridges and buildings, the common practice in the past was designed on the basis of a comparison of allowable stress intensities with those produced by the service loadings and other design conditions. This traditional design is called elastic design because the allowable stress intensities are chosen according to the concept that the stress or strain corresponding to the yield point of the material should not be exceeded at the highest stressed points of the structure. The selection of the allowable stresses may also be modified by consideration of the possibility of failure due to fatigue, buckling, or brittle fracture or by consideration of the permissible deflections of the structure. Now, *limit state design*[2] has replaced the older concept of permissible stress design in most forms of civil engineering.

Engineers often evaluate structural loads based upon published regulations, contracts, or specifications. Accepted technical standards are used in acceptance testing and inspection.

2.3.2 Notes

1. Creep and shrinkage of concrete are two physical properties of concrete. The creep of concrete results from the calcium silicate hydrates (CSH) in the hardened Portland cement paste (which is the binder of mineral aggregates). Changes of pore water content due to drying or wetting processes result in significant volume changes of concrete in load-free specimens. They are called shrinkage (typically causing strains between 0.000 2 and 0.000 5, and in low strength concretes even 0.001 2) or swelling (< 0.000 05 in normal concretes, < 0.000 20 in high strength concretes).

混凝土的徐变和收缩是混凝土的两个物理性质。混凝土的徐变来源于硬化波特兰水泥浆（这是矿质集料的黏结剂）中的硅酸钙水合物（CSH）。在干燥或润湿过程中孔隙水含量的变化会导致无载荷试样中混凝土体积发生显著变化。它们被称为收缩（通常在 0.000 2 至 0.000 5 之间产生应变，在低强度混凝土中甚至为 0.001 2）或膨胀（在普通混凝土中为 0.000 05，在高强度混凝土中<0.000 20）。

2. Limit state design (LSD), also known as load and resistance factor design (LRFD), refers to a design method used in structural engineering. A limit state is a condition of a structure beyond which it no longer fulfills the relevant design criteria.

极限状态设计（LSD）又称载荷阻力系数设计（LRFD），是指结构工程中使用的一种设计方法，极限状态是结构超出限制而不再满足相关设计标准的一种情况。

2.3.3 New Words and Expressions

structural design	结构设计
engineering project	工程项目
planning	规划
design	设计
construction	施工
dimension	n.尺寸；【数】维；v.标出尺寸
structural element	结构构件
mathematical modeling	数学模型
structural engineering	结构工程
the preliminary planning	初步规划
permanent load	永久负载，恒载荷
magnitude ['mægnɪtju:d]	巨大，广大；重大，重要；量级；（地震）级数
variable load	可变载荷
dead load	永久固定的载荷（如房屋、桥梁等）；自重；恒载荷
self weight of the structure	结构自重
live (or imposed) load	活载荷

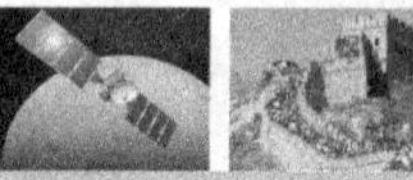

wind load	风力载荷；风载荷
windward [ˈwɪndwəd]	adj.& adv.迎风的（地）；n.上风，迎风
suction [ˈsʌkʃn]	吸，抽吸
leeward [ˈli:wəd]	adj.& adv.背风的（地），下风的（地）；n.下风
impact load	冲击负载
applied load	外加负载，施加载荷
fatigue	疲劳
buckling	弯折，压曲
brittle-fracture	脆裂
yield	屈服
elastic-plastic displacement	弹塑性位移
the point of collapse	破坏点
state of equilibrium	静力平衡
the external load	外加载荷
supporting force	支承力
reaction	反作用力
stress	应力强度
deformation	变形
displacement	位移
structural analysis	结构分析
excess load	超载荷
overloading	超载
structure failure	结构失效破坏
internal gross forces	总内力
thrust [θrʌst]	推力
shear[ʃɪə(r)]	剪力
bending moment	弯矩
twisting moment	扭矩
stress intensity	应力强度
strain[streɪn]	应变
deflection [dɪˈflekʃn]	偏斜；偏斜度，挠度
shrinkage [ˈʃrɪŋkɪdʒ]	收缩
creep	徐变
allowable stress intensities	允许应力强度
service loading	使用载荷，工作载荷
elastic design	弹性设计
yield point	屈服点

limit state design	极限状态设计
code	规范
regulation [ˌregjuˈleɪʃn]	规则
contract [ˈkɒntrækt]	合同
specification [ˌspesɪfɪˈkeɪʃn]	规范

2.4 Bidding Document and Contract 标书与合同

扫一扫看本节参考译文

2.4.1 Text

扫一扫看本节教学课件

The *bidding documents*[1] should illustrate clearly whether contracts will be awarded according to *unit prices*[2] or a *lump sum*[3] in the contract, on the basis of the nature of goods or works to be provided.

The size and scope of contracts will depend on the magnitude, nature, and location of the project. For projects requiring various works and equipment such as power, water supply, or other industrial projects, separate contracts are commonly awarded for the civil works, and for the supply and erection of different major items of plant and equipment. *Contractor*[4]s or manufacturers, small and large, should be allowed to bid for individual contracts or for a group of similar contracts at their option, and all bids and combinations of bids should be opened and evaluated so as to determine each bid or combination of bids offering the best solution for the borrower.

Detailed civil engineering civil works, including the preparation of technical specifications and other bidding documents, should precede the invitation to bid for the contract. Nevertheless, in the case of *turnkey contract*[5]s or contracts for large complex industrial projects, it may be undesirable to prepare technical specifications in advance. In such a case, it will be necessary to use a two-step procedure to invite un-priced technical bids so as to technical clarification and adjustments, and then to submit price proposals.

The preparation time for bids should depend on the magnitude and complexity of the contract. Generally, international bidding needs no less than 45 days from the date of bidding invitation. However, large civil works are involved no less than 90 days from the date of invitation to enable prospective bidders to conduct investigations at the site before submitting their bids. The time allowed should be governed by the particular circumstances of the project.

The expiration date, hour, and place of bids and the bid opening, should be announced in the bidding invitation. All bids should be opened at the stipulated time. Bids delivered after the time stipulated should be returned unopened unless the delay fault was not due to the bidder and its late acceptance would not give him any advantage over other bids. Bids should be opened in public. The name of the bidder and total amount of each bid should be read aloud and recorded when it is opened.

Extension of validity of bids should commonly not be requested. Bidders should have the right to refuse to grant such an extension without forfeiting their bid bond, but those who agree to extend the validity of their bid should be neither required nor permitted to modify their bids.

It is inadmissible that information relating to the examination, clarification, and evaluation of bids and recommendations concerning awards is revealed to bidders and persons who have no official relations with these procedures until the award of a contract.

No bidder should be permitted to alter his bid after bid has been opened. Only clarifications without changing the actual substance of the bid may be accepted. The borrower may ask any bidder for a clarification of his bid but should not ask any bidder to change the actual substance or price of his bid.

A detailed report on the evaluation and comparison of bid, which sets forth the specific reasons and points successful bidder out clearly, should be prepared by the borrower or by its consultants.

The award of a contract should be made within set time and be awarded to the bidder whose bid has been determined to be lowest evaluated bid and applies for the appropriate standards in the aspects of capability and financial resources.

2.4.2 Notes

1. Bidding is an offer (often competitive) to set a price by an individual or business for a product or service or a demand that something be done. Bidding is used to determine the cost or value of something.

投标是一种由个人或企业为某一产品或服务设定价格的要约（通常是具有竞争性的），或者是一种要求做某事的要约。投标是用来确定某物的成本或价值的。

2. Unit price is the price of every piece of commodity.

单价是每单位商品的价格。

3. lump sum: an amount of money that is paid at one time and not on separate occasions.

一次性付的总钱款。

4. contractor: a person or company that has a contract to do work or provide goods or services for another company.

承包人；承包商；承包公司。

5. A turnkey or a turnkey project is a type of project that is constructed so that it can be sold to any buyer as a completed product.

交钥匙或交钥匙项目是一种工程，其建造方式是可以将其作为已完成的产品出售给任何买方。

Build to order (BTO) and sometimes referred to as make to order or made to order (MTO), is a production approach where products are not built until a confirmed order for products is received.

生产到订单（BTO），有时被称为“按订单制造”（MTO），是指在收到确认的产品订单之前不生产产品的生产方法。

Turnkey contract: 包括规划、设计和管理内容的施工合同，为整套承包合同。

2.4.3 New Words and Expressions

tender document	投标文件
bidding document	招标文件
award	授予
unit price	单价，分项价格
lump sum	总价格，总金额，包干价
nature	性质
civil works	土木工程，土建
erection [ɪ'rekʃn]	安装，装配，建设
bid	出价，投标
contractor	承包人
manufacturer [ˌmænju'fæktʃərə]	制造商
combination of bids	组标
borrower	借款人
turnkey['tɜ:nki:]	包到底的工程，交钥匙工程
turnkey contract	包括规划、设计和管理的施工合同，整套承包合同
technical specification	技术规范
un-priced technical bid	不确定费用的技术招标
international bidding	国际投标
clarification [ˌklærəfɪ'keɪʃn]	说明
submit [səb'mɪt]	呈报，提交
stipulate ['stɪpjuleɪt]	（尤指在协议或建议中）规定，约定，讲明（条件等）
bidder ['bɪdə(r)]	出价者，投标人
extension [ɪk'stenʃn]	延期
validity [və'lɪdəti]	有效，合法性；效力
expiration [ˌekspə'reɪʃn]	截止日期
grant[grɑ:nt]	承认；同意；准许；授予
forfeit ['fɔ:fɪt]	（因违反协议、犯规、受罚等）丧失，失去
bid bond [bid bɔnd]	投标保证；履约担保书；投标保证金；押标金；
examination[ɪgˌzæmɪ'neɪʃn]	审查，检查
evaluation [ɪˌvælju'eɪʃn]	评价
consultant[kən'sʌltənt]	顾问

知识分布网络

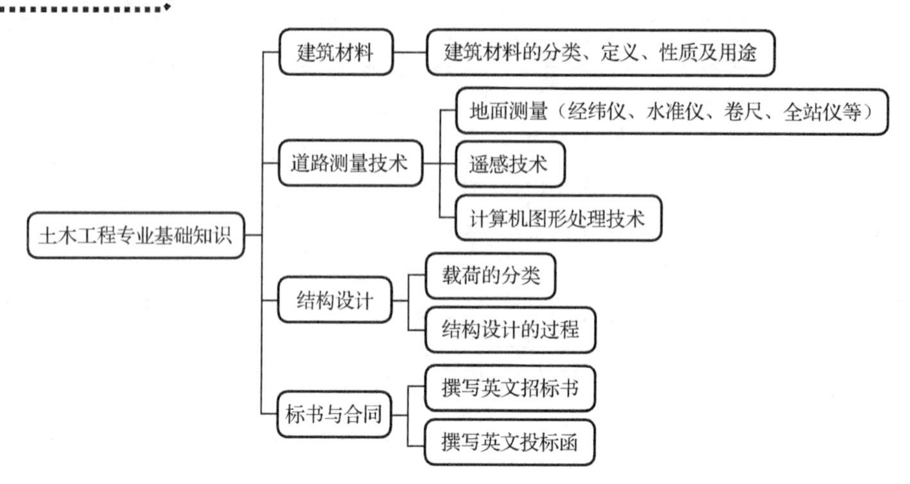

知识梳理与总结

通过对本章的学习，学生能够明确学习目标，依据教材、词典、多媒体课件、视频等学习手段和工具掌握建筑材料、道路测量技术、结构设计、标书与合同等土木工程专业基础的英语知识，并且熟悉、背诵相关的专业英语词汇。本章必须掌握的知识要点有：

1. 背会建筑材料的专业英语词汇，并且会用英文表达主要建筑材料的定义、成分、性质及用途，如沥青、水泥、钢筋、钢筋混凝土等；
2. 背会道路测量技术的专业英语词汇，会使用英语进行角度测量、距离测量等的表述；
3. 背会结构设计的专业英语词汇；
4. 背会工程标书与合同的专业英语词汇，并且能够初步运用英语撰写招标书、投标函。

希望学生能够阅读与土木工程专业相关的英文文献、学术论文等并且撰写英文标书与合同，并自我培养专业英语的阅读及表达能力。

思考与练习题 2

扫一扫看第 2 章习题答案

Task 1: Match the words given in Column A with the meanings given in Column B.

A	B
1. aggregate	a. award; give a formal or legal permission
2. asphalt	b. equipment used by surveyors for measuring angles
3. cement	c. weight
4. reinforced concrete	d. measurement
5. theodolite	e. pay for a price, in competition with other companies
6. level	f. the state of being legally or officially acceptable
7. tape	g. an amount of money that is paid at one time
8. total station	h. measure by taking a picture

9. remote sensing	i. a grey powder made by burning clay and lime
10. load	j. a person or company that has a contract to do work
11. bid	k. deadline
12. lump sum	l. concrete with metal bars inside to make it stronger
13. surveying	m. bitumen
14. photogrammetry	n. sand, stone used to make concrete or build roads
15. consultant	o. measure distance and elevation by airplane
16. contractor	p. changing and spoiling the normal shape
17. expiration	q. test whether surface is level
18. validity	r. used for direct measurement of horizontal distance
19. grant	s. determine angles and distance from the instrument
20. deformation	t. a person employed to give advice to other people

Task 2: Choose one correct answer from A, B, C and D.

1. (　　) surveying-defined as the use of electronic survey equipment used to perform horizontal and vertical measurements in reference to a grid system.

A. Theodolite　　B. spirit level　　C. Transit　　D. Total station

2. A total station integrates the functions of a theodolite for measuring angles, an (　　) for measuring distances, digital data and a data recorder.

A. EDM　　B. DTM　　C.IMU　　D. EDI

3. Typical total station programs include point (　　),missing line measurement, azimuth calculations, remote object elevation calculations, offset measurements, layout or setting-out positions, and area computation.

A. level　　B. horizontal　　C. height　　D.location

4. The primary function of surveying instruments is to measure distances, angles and (　　).

A. length　　B. radius　　C. position　　D. heights

5. Level is a device for establishing a horizontal line or plane by means of a (　　) in a liquid that shows adjustment to the horizon by movement to the center of a slightly bowed glass tube.

A. tube　　B. cross hair　　C. bubble　　D. level bar

6. The instrument that can not directly measure angles is (　　).

A. theodolite　　B. taping　　C. level　　D. clinometer

Task 3: Translate the following into Chinese.

1. Reinforced concrete (RC) is a composite material in which relatively low tensile strength and ductility of concrete are counteracted by the inclusion of reinforcement with higher tensile strength or ductility.

2. Detailed civil engineering civil works, including the preparation of technical specifications and other bidding documents, should precede the invitation to bid for the contract.

3. Now, limit state design has replaced the older concept of permissible stress design in most forms of civil engineering.

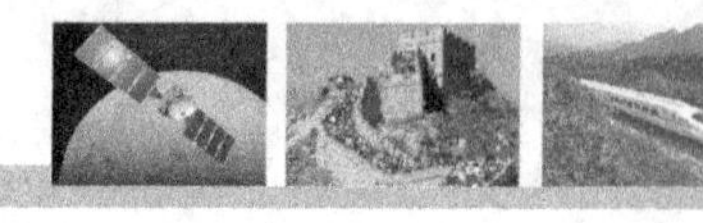

4. An electronic distance measurement (EDM) device consists mainly of a transmitter located at one end of the distance to be measured and a reflector at the other end.

5. Loads may be broadly classified as permanent loads that are constant in magnitude and remain in one position and variable loads that may change in position and magnitude.

Task 4: Fill in the blanks according to the Chinese given.

中标通知书

合同号：2011

日期：2022 年 5 月 13 日

致：哈尔滨第一建筑公司（中标人名称和地址）

先生们：

贵单位于 2022 年 4 月 15 日为建设大庆团结桥（工程名称）以人民币 752 000 元所提交的投标书已被我方接受。请做好签署合同的准备。谨致。

张鹏飞（被授权代表甲方签署本通知书的人签字并加盖公章）

2022 年 5 月 13 日

A Letter of Acceptance（中标通知书）

Contract NO. ____1____

Date: ____2____

To: ________3________ (name and address of the successful bidder)

Dear Sir,

This is to notify you that the works of your____4____(bid date) for construction of ____5____(name of project) for the contract price of____6____Yuan (RMB) is hereby accepted by our agency. You are asked to be ready for signing the contract. Yours Sincerely.

____7____(signature, name, and title of signatory authorized to sign on behalf of the employer).

Date: May, 13th, 2022

Chapter 3 Highway Engineering

第3章　道路工程

教学导航

教	知识重点	3.1 General Introduction of Highway 公路概况; 3.2 Highway Design 公路设计; 3.3 Subgrade Engineering 路基工程; 3.4 Pavement Engineering 路面工程; 3.5 Highway Interchange 公路互通式立体交叉; 3.6 Highway Maintenance & Management 道路养护与管理
	知识难点	道路工程相关专业英语词汇的背诵与记忆
	推荐教学方式	明确教学目标，依据教材、多媒体课件、视频等教学手段让学生掌握道路工程相关专业英语知识并且利用科学的单词记忆法引领学生背诵、积累道路工程领域的专业英语词汇
	建议学时	16学时
学	推荐学习方法	以任务驱动、小组讨论、课外拓展的学习方式为主。结合本章内容，课堂学习与自主学习相结合
	必须掌握的理论知识	道路工程相关专业英语知识与词汇
	必须掌握的技能	通过对本章节的学习，学生能够识读道路工程相关专业英语词汇并能够阅读与道路工程相关的英文文献、学术论文等。

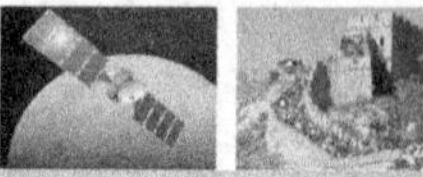

3.1 General Introduction of Highway 公路概况

扫一扫看本节参考译文

扫一扫看本节教学课件

3.1.1 Text

A *highway*[1] is any public *road*[2] on land (Fig.3-1). It is used for major roads, but also includes other public roads and public tracks by people on foot or on horses. Later they also held carriages, bicycles and eventually motor cars, promoted by advancements in road construction. In the 1920s and 1930s, many countries began investing heavily in progressively modern highway systems to spur commerce and bolster national defense. It is not an equivalent term to controlled-access highway, or a translation for autobahn, autoroute, etc.

Fig.3-1 The term highway includes any public road. This is an unpaved highway in Heilongjiang Province

History of Highway

The first road builders in Western Europe were the Romans. Roman roads are characterized by their linearity and durability. A good alignment was the most direct route gained after reducing the risk of ambush in hostile territory. It was for this reason that the surface of the road was often elevated a meter or more above the local ground level so as to provide a clear view of the surrounding country; producing the modern term "highway".

A typical major Roman road consisted of several layers of material, increasing in strength from the bottom layer of rubble, through intermediate layers of lime-bound concrete to an upper layer of flags or stone slabs grouted in lime (Fig.3-2).

Fig.3-2 Road Ruins of Ancient Rome

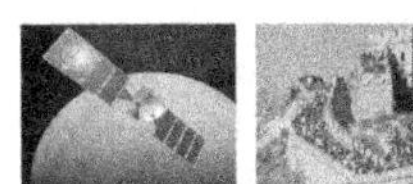

Modern highway systems developed in the 20th century as the automobile became popularity. The world's first limited access road was constructed on Long Island New York in the United States known as the Long Island Motor Parkway. It was finished in 1911. Construction of the Bonn–Cologne autobahn in Germany began in 1929 and it was opened in 1932 (Fig.3-3). In China, the term "highway" firstly appeared in 1920s, when the Highway Department was founded in Guangdong Province.

Fig.3-3　A German Autobahn in the 1930s

At present, Australia's Highway 1 is the longest national highway in the world at over 14 500 km or 9 000 mi and runs almost the entire way around the continent. China has the world's largest network of highways followed closely by the United States of America.

Components of Highway

Highway is a linear structure to support vehicle load. It basally includes supporting structure (roadbed or subgrade, pavement), bridge, culvert, tunnel, protective work, drainage system, intersection/junction/crossing, interchange, toll road, traffic service facility and so on.

Subgrade is a kind of structure which is excavated or filled according to the highway alignment design and technical requirement. It mainly supports the vehicle load transiting from the surface, and it is a foundation to support the pavement. Figure3-4 depicts the cross section of the roadbed. When designed, subgrade should have sufficient strength and stability and prevent from the damage of water and other natural factors.

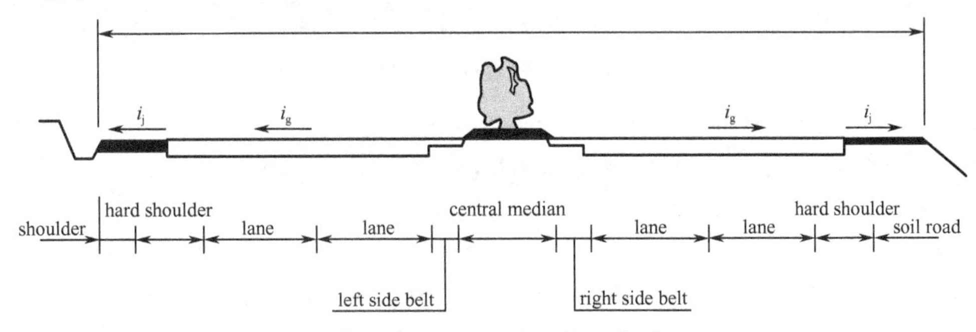

Fig.3-4　Cross section of roadbed

Subgrade is usually compacted before the construction of a road, pavement or railway track, and it is sometimes stabilized by the addition of asphalt, lime, Portland cement or other modifiers. The subgrade is the foundation of the pavement structure, on which the subbase is laid (Fig.3-5).

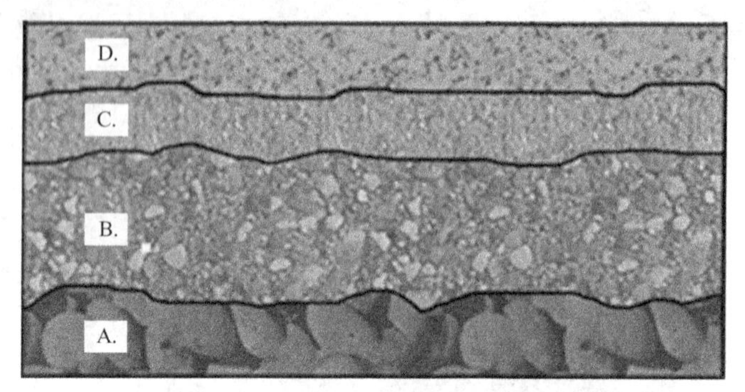

Fig.3-5 Layers in the construction of pavement

(A. Subgrade B. Functional layer C. Subbase D. Base course)

The load-bearing strength of subgrade is measured by California Bearing Ratio (CBR) test, falling weight deflectometer back calculations and other methods.

Pavement is made up of various materials placed in layers (Fig.3-5) and used to offer traveling for vehicles. It can be divided into two categories according to paver materials, namely rigid pavement and flexible pavement (Fig.3-6). The cement road represents rigid pavement and the asphalt pavement stands for flexible pavement. In designing, the pavement should ensure enough strength, rigidity, smoothness and roughness to meet the need that vehicles can travel safely, fast and comfortably on it.

Fig.3-6 Rigid pavement and flexible pavement

Bridge is constructed for the road to traverse over obstacles such as rivers, valleys, and artificial structures. *Culvert*[3] is a structure that allows water to flow under a road, railroad, trail, or similar obstruction from one side to the other side (Fig.3-7).

Tunnel is built as the design requirement for the road to traverse through mountains, underground or water bottom. It is an underground passageway, dug through the surrounding soil/earth/rock and enclosed except for entrance and exit, commonly at each end.

Fig.3-7　Culvert

Drainage system is part of highway for discharging the surface water and ground water. The system mainly contains side ditch, drainage ditch, blind ditch, leak ditch, drain pipe and so on (Fig.3-8).

Fig.3-8　Side ditch

Protective work is a structure erected for reinforcing the side slope and ensuring the stability of roadbed, generally including plant and engineering protection (Fig.3-9).

Fig.3-9　Plant and engineering protection

Intersection is a road cross at grade. Interchange is a road junction that typically uses grade separation with one or more ramps, to permit traffic on at least one highway to pass through the junction without directly crossing any other traffic stream (Fig.3-10).

Fig.3-10 Interchange

Traffic service facility is the facility arranged along the routes such as traffic sign, marking, guardrail, central median, noise barrier, lighting and landscaping facility.

Types of Highway

Highways and streets are categorized as rural or urban roads, depending on the area in which they are located. more over, Highways are classified into the following groups according to the service functions provided:

- Arterial highway—A general term denoting a highway primarily for through traffic on a continuous route, or a large and important road, river, railway/railroad line, etc.
- Minor arterial highway—A regional arterial highway.
- Bypass—An arterial highway that permits traffic to avoid part or all of an urban area.
- Local roads and streets—A street or road primarily for access to residence, business, or other abutting property.

Freeways are not listed as a separate functional class since they are generally classified as part of the principal arterial system.

3.1.2 Notes

1. highway 公路是公共道路的简称，习惯上指各级政府所建的连接城市和乡镇之间的、具有一定技术标准和设施配置的道路。

2. road 道路是指能够通行的途径。不同的行为主体，对道路的界定标准不同。本领域引用的道路概念，一般是指机动车辆和行人均能通行的途径。

3. If the span of crossing is greater than 12 feet (3.7 m), the structure is termed as bridge and otherwise is culvert.

如果跨度大于 12 英尺（3.7 米），则该结构称为桥梁，否则称为涵洞。

3.1.3 New Words and Expressions

major road	主干道，主要道路
track	小路，小道
equivalent[ɪˈkwɪvələnt]	相等的，相当的
controlled-access highway	限制进入的道路，快速道
autobahn [ˈɒtəʊba:n]	（德国、奥地利或瑞士的）高速公路
autoroute [ɔ:təʊ'ru:t]	（法国和法语地区的）高速公路
linearity [ˌlɪnɪ'ærətɪ]	线性，直线性
durability [ˌdjʊərə'bɪlətɪ]	耐久性
alignment [əˈlaɪnmənt]	排成直线
route [ru:t]	路，路线
ambush[ˈæmbʊʃ]	伏击，埋伏
hostile[ˈhɒstaɪl]	敌人的，敌方的
territory[ˈterətri]	领土，领地
rubble [ˈrʌbl]	块石，瓦砾
lime-bound concrete	石灰混凝土结合料
flag	薄层，薄层砂岩
slab [slæb]	平板，混凝土路面
lime [laɪm]	石灰
automobile [ˈɔ:təməbi:l]	汽车
limited access road	限制进入的道路，快速道
motorway	（英国）高速公路
parkway	a wide road with trees，（有草木的）大路
Bonn [bɔn, bɔ:n]	波恩（德国城市）
Cologne [kəˈləʊn]	科隆（德国城市）
linear　[ˈlɪniə(r)]	线性的，直线的
vehicle [ˈvi:əkl]	车辆，交通工具
roadbed	路基，路床
subgrade	路基
pavement	路面
culvert[ˈkʌlvət]	涵洞

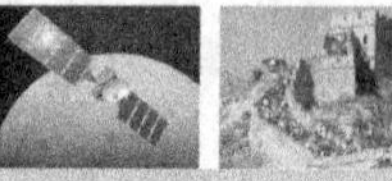

tunnel	隧道
protection work	防护工程
drainage [ˈdreɪnɪdʒ]	排水系统
intersection	相交
junction[ˈdʒʌŋkʃn]	（公路的）交叉路口
crossing	交叉路口，十字路口
interchange	互通式立体交叉道
toll road	收费公路
traffic service	交通服务
facility	设施
excavate[ˈekskəveɪt]	挖凿，开挖
fill	填方，填
highway alignment design	公路线型设计
foundation	基础
depict	描述
cross section	横断面，横截面
strength	强度
stability	稳定性
shoulder	路肩
hard shoulder	硬路肩
lane	行车道
central median	中央分隔带
side belt	路缘带
compaction	压实
modifier	改性剂
subbase	底基层
base course	基层
load-bearing strength	承载强度
California Bearing Ratio (CBR)	加州承载比
falling weight deflectometer	落锤式弯沉仪
rigid pavement	刚性路面
flexible pavement	柔性路面
obstacle	障碍（物）
obstruction	障碍物
discharge	排放

side ditch	边沟
drainage ditch	排水沟
blind ditch	盲沟
leak ditch	渗沟
rain pipe	排（雨）水管
reinforce	加固，加强
side slope	边坡
at grade	在同一水平面上
grade separation/ intersection	立体/平面交叉
ramp [ræmp]	匝道
traffic sign	交通标志
marking	标记
guardrail [ˈgɑːdreɪl]	护栏
noise barrier	噪声屏障
lighting facility	照明设施
landscaping facility	绿化景观设施
rural/ urban road	乡村/城市道路
arterial [ɑːˈtɪərɪəl]	主干道，干线（指主要公路等）
bypass	（绕过城市的）旁路，旁道，支路

3.2　Highway Design 公路设计

扫一扫看
本节参考
译文

扫一扫看
本节教学
课件

3.2.1　Text

The Concept and Philosophy of Highway Design

Design Objectives　Highway geometric design refers to the calculation and analysis by transportation engineers (or designers) to make the highway fit the topography of the site, while the design should meet the safety, service and performance standards. It mainly involves the elements of the highways that are visible to the drivers and users. However, the engineer must also take into consideration the social and environmental impacts of the highway geometry on the surrounding facilities.

Usually, highway geometric design has the following objectives:

(1) With the allowance of the design standard and right-of-way, to determine the proposed routing of highway.
(2) According to the design standard, to incorporate physical features of the road alignment to ensure that drivers have sufficient view of the road ahead for them to adjust their speed of traveling to maintain safety and ride quality.
(3) To provide a basis for the highway engineers to evaluate and plan for the construction of a section of the proposed highway.

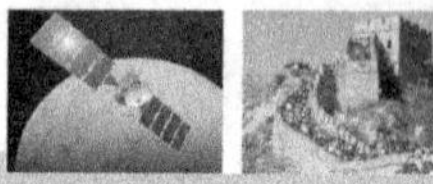

Design Considerations To make the highway suitable for the topography of the construction site and meet the requirements of the safety, service and performance standards, the following factors have to be taken into consideration daring the design process:

Design speed	Design traffic volume	Number of lanes	Level of service	Sight distance
Alignment, super-elevation and grades	Cross section	Lane width	Horizontal clearance	vertical clearance

Design Process A highway designer should pay attention to four major fields perticalerly during the different periods of the project planning and design phases: location design, alignment design, cross sectional design, and access design.

Location design takes place at the earlier stage of project planning. It refers to the macro-level routing of a planned highway connecting two points through the existing highways, communities and natural terrain. Then the designer goes through the various alterative and consultative steps with the stake holders to modify and select the most feasible layout. The consultative process may take several months. A more detailed ground survey map is then carried out to locate the key control points of the alignments according to geo-coordinates and elevations. Finally, the designer proceeds with the detailed alignment cross sectional and assess design.

Highway Alignment Design

The alignment design of roads is the branch of highway engineering concerned with the positioning of the roadway. The alignment of a highway is a three-dimensional problem because the highway itself travels over the terrain to connect two points. The highway may be visualized as segments of connected horizontal and vertical curves (or their combination). The alignment of a highway is represented by its centre line in a three-dimensional coordinate system (e. g. longitude, latitude, and elevation).

three main parts of Geometric roadway design can be separated as alignment, profile, and cross-section. Combined, they provide a three-dimensional layout for a roadway:

alignment	The route of the road, defined as a series of horizontal tangents and curves
profile	The vertical aspect of the road, including crest and sag curves, and the straight grade lines connecting them
cross section	Cross section shows the position and number of vehicle and bicycle lanes and sidewalks, along with their cross slope or banking. Cross sections also show drainage features, pavement structure and other items outside the category of geometric design

Horizontal Alignment Horizontal alignment in road design consists of straight sections of road, known as tangents, connected by circular horizontal curves. Circular curves are defined by radius (tightness) and deflection angle (extent). Horizontal curves are expressed as circular curves with constant radii, or successive curves with different radii (Fig.3-11). A curve can be described by its radius or by its degree of curvature. The design of a horizontal curve entails the determination of a minimum radius (based on speed limit), curve length, and objects obstructing the view of the driver.

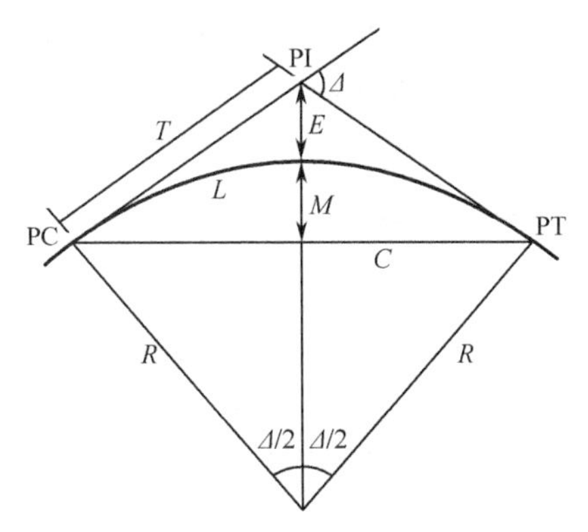

Fig.3-11　Horizontal curve[1]

Using *AASHTO*[2] standards, an engineer works to design a road that is safe and comfortable. If a horizontal curve has a high speed and a small radius, an increased superelevation is needed in order to assure safety. If there is an object obstructing the view around a corner or curve, the engineer must work to ensure that drivers can see far enough to stop in time to avoid an accident or accelerate to join traffic.

Figure 3-11 shows the properties of a curve with a constant radius (R) connecting two straight sections of a highway. The curve starts at point of curvature (PC), ends at point of tangent (PT). The point of intersection (PI) is the intersecting point if the two straight lines are extended. Δ is the central angle of the curve, expressed in degrees.

The length of tangent (T) is

$$T = R\tan\left(\frac{\Delta}{2}\right)$$

The external distance (E) is

$$E = R\left(\frac{1}{\cos\left(\frac{\Delta}{2}\right)} - 1\right)$$

Vertical Alignment The profile grade line defines the vertical alignment for construction in terms of straight grades and parabolic curves. The characteristics of vertical alignment are influenced greatly by basic controls related to design speed, traffic volume functional classification and terrain conditions. Major criteria for profile consist of maximum grades, minimum grades, minimum ditch grades, critical length of grade, vertical curves and gradeline elevations.

A vertical curve is defined as a smooth transition curve between two tangent grades so as to provide a safe, comfortable sight distance. The principal concern in designing vertical curves is to ensure that at least the minimum stopping sight distance is provided. There are two types of vertical curves: crest vertical curves and sag vertical curves (Fig.3-12). As a departure from the horizontal

curve, the points of curvature, intersection and tangent of a vertical curve are denoted by PVC, PVI and PVT, respectively. The length *L* of curve is the distance between PVC and PVT measured along the horizontal plane. The PVI is at the midpoint between PVC and PVT along the horizontal plane.

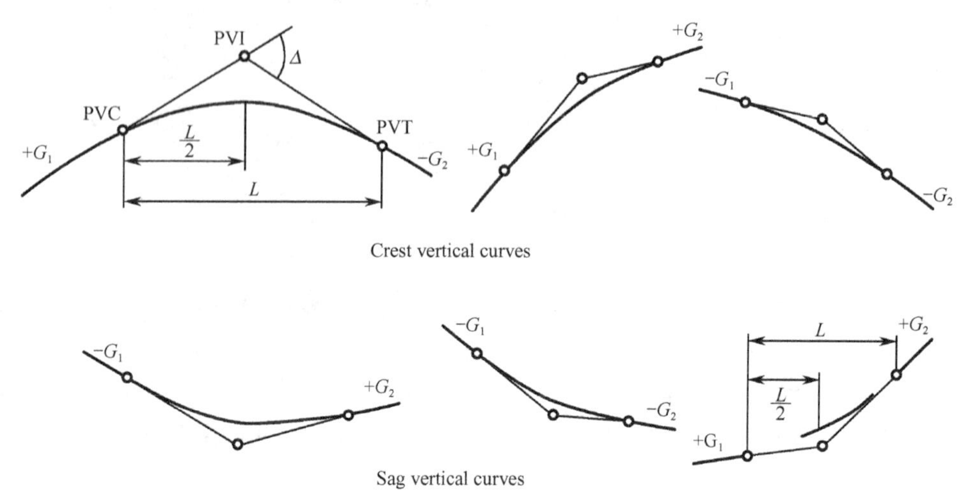

Fig.3-12 Crest vertical curves and sag vertical curves[3]

General criteria 1. Use a smooth grade line with gradual changes, consistent with the type of highway and character of terrain, rather than a line with numerous breaks and short lengths of tangent grades.

2. Avoid very long crest vertical curves if passing sight distance cannot reasonably be attained. A shorter vertical curve may permit more passing opportunity on adjacent tangent grades.

3. On a long grade it is preferable to place the steepest grade at the bottom and to flatten the grade near the top.

4. Both horizontal curvature and the profile should be as flat as feasible at intersections where sight distances along both roads and streets are important and vehicles may have to slow or stop. Maintain moderate grades through intersections to facilitate turning movements.

5. Consider auxiliary lanes where passing opportunities are limited and it is probable that slow-moving vehicles will affect operating speeds and the desired level of service.

6. Sharp horizontal curvature should not be introduced near or just beyond the top of a pronounced crest vertical curve. This condition makes it difficult for drivers to perceive the horizontal changes in alignment, especially at night.

7. Sharp horizontal curvature should not be introduced at or near the low point of a pronounced sag vertical curve.

8. On two-lane roads and streets with considerable traffic volume, safe passing sections must be provided at frequent intervals and for an appreciable percentage of the length of roadway. In these cases, it is necessary to work toward long tangent sections to secure sufficient passing sight distance rather than the more economical combination of vertical and horizontal alignment (Fig.3-13).

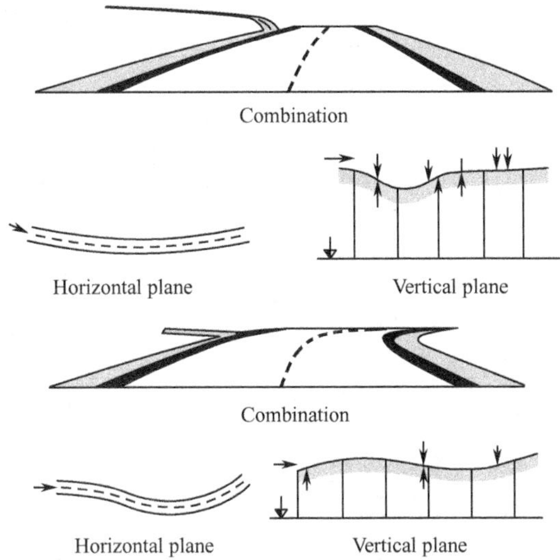

Fig.3-13　The combination diagram of horizontal curve and vertical curve

3.2.2　Notes

1. Terminology 术语

R—Radius.

PC—Point of Curvature (point at which the curve begins).

PT—Point of Tangent (point at which the curve ends).

PI—Point of Intersection (point at which the two tangents intersect).

T—Tangent Length.

C—Long Chord Length (straight line between PC and PT).

L—Curve Length.

M—Middle Ordinate, now known as HSO (Horizontal Sightline Offset, distance from sight-obstructing object to the middle of the outside lane).

E—External Distance.

u—Vehicle Speed.

Δ—Deflection Angle.

2. AASHTO: American Association of State Highway and Transportation Officials，美国国有公路运输管理员协会。

3. G_1—the initial grade.

G_2—final grade.

3.2.3　New Words and Expressions

highway geometric design	公路几何设计
calculation	计算
transportation engineer	交通工程师
topography[tə'pɒgrəfi]	地形学

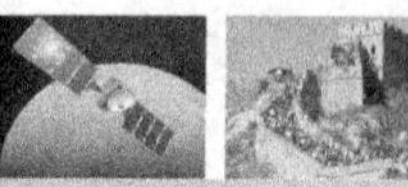

geometry[dʒiˈɒmətri]	几何学
right-of-way	公路用地
traffic volume	交通量
sight distance	视距
section	部分，断片，截面
grade	坡度
cross section	横断面
clearance	净空
location design	选线设计，定位设计，线路设计
alignment design	线形设计
cross section design	横断面设计
access design	路口设计
project planning	项目规划
macro-level	宏观层面
terrain[təˈreɪn]	地形，地势；地面，地带；［地理］岩层
stake holder	股份持有人
feasible [ˈfi:zəbl]	可行的
layout	布局
ground surevey map	勘测图
roadway	道路，车道
three-dimensional	三维的
segment[ˈsegmənt]	部分、段落；分割
horizontal curve	平面曲线，水平曲线
vertical curve	竖曲线
centre line	中心线
three-dimensional coordinate system	三维坐标系统
longitude [ˈlɒŋgɪtju:d]	经度
latitude [ˈlætɪtju:d]	纬度
profile [ˈprəʊfaɪl]	剖面
tangent [ˈtændʒənt]	【数】正切；（铁路或道路的）直线区间
crest vertical curve	凸形竖曲线
sag vertical curve	凹形竖曲线
straight grade line	直坡线
sidewalk	人行道
drainage [ˈdreɪnɪdʒ]	排水，排水系统
circular curve	圆曲线
radius[ˈreɪdiəs]	半径

deflection angle	偏转角，偏角
extent	程度
radii['reɪdɪaɪ]	半径；半径（距离）（radius 的名词复数）
degree of curvature	弯曲度，曲度
curvature [ˈkɜ:vətʃə(r)]	曲率
entail[ɪnˈteɪl]	牵涉；需要
speed limit	速度限制
curve length	曲线长度
superelevation	（公路的）超高（公路转弯处外侧比内侧高出的程度）
grade line	纵坡线，坡度线
parabolic[ˌpærəˈbɒlɪk]	抛物线的
parabolic curve	抛物曲线
maximum grade	最大纵坡
minimum grade	最小纵坡
minimum ditch grade	最小边沟纵坡
critical length of grade	边沟的临界长度
gradeline elevation	纵坡线的高程
grade	纵坡
tangent grade	直坡线，直坡段
auxiliary lane	辅道
two-lane road	双车道道路

3.3 Subgrade Engineering 路基工程

扫一扫看本节参考译文

扫一扫看本节教学课件

3.3.1 Text

The Subgrade Design

Highway Subgrade (or basement soil) is defined as the supporting structure of the layers of pavement. In cut section, the subgrade is the original soil lying below the layers designated as base and subbase. In fill sections, the subgrade is constructed over the nation ground with imported material from nearby roadway cuts or from borrow pit and compacted to a specified density and moisture content.

The cross-sectional shape of the subgrade depends on the type of the surface of the earth (Fig.3-14).

Before 1920, the pavement or other wearing courses were paid more attention on. Conversely, the materials of the subgrade and its filling and compacting manners were paid little attention on. Soon after, increasing speed of vehicle brought demands for higher standards of alignment and stability, which in turn meant deeper cuts and higher fills.

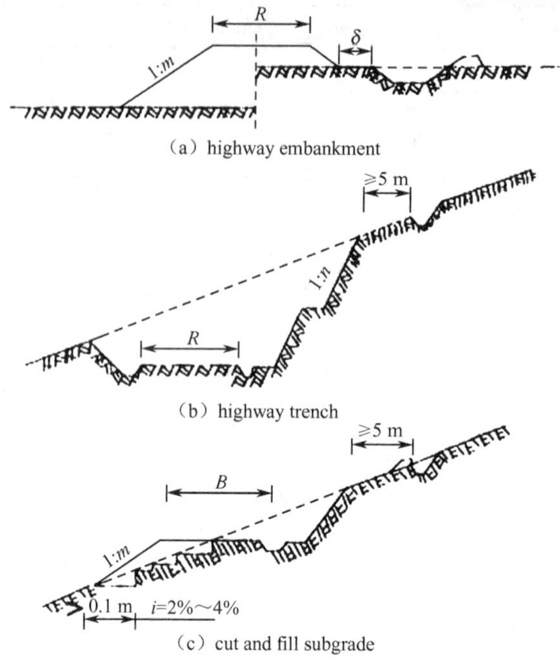

Fig.3-14 Types of Highway Subgrade

About the same route, the weight and number of vehicles began increasing, which imposed larger and more numerous wheel loads on the roadway surface. In many instances, subsidence or even total failure of the roadway appeared. Study of such failures indicated that the fault resulted from the subgrade and not from the pavement. So, the properties of subgrade soil and of its function under service conditions should be investigated.

1. *Soil Classification*[1]

The basic components of soils are simply classified on the basis of grain size as follows:

giant grained soil	cobbles	>75 mm
coarse grained soil	gravel	75 mm~5 mm
	sand	5 mm~0.075mm
fine grained soil	silt	0.075 mm~0.002 mm
	clay	< 0.002 mm

Fine grained soils are defined as materials with more than 50 percent of the mass smaller than the 0.075 mm particle size. It is necessary to understand that plasticity is an extremely important property to differentiate between silt and clay. The subgrade design engineers are most interested in the strength of the soil and the extent to which this strength varies with climate, environment and drainage effects. The properties of a soil mixture are influenced more by moisture than by any other cause. Soils that have sufficient strength and supporting power under a certain moisture condition may is completely at odds if the percentage of moisture changes. One difficulty with soil in highway subgrade is that they are susceptible to such moisture changes.

The physical properties of soil consist of moisture content, density, liquid and plastic limit:

moisture content	Moisture content is defined as the ratio of the mass of water present in a body of soil to the mass of the dry soil particles. It is often expressed as a percentage and is measured by weighing a sample of the soil, drying it—usually in an oven at a temperature of 105℃, and weighing it again. The mass of water is determined by subtraction.
density of a soil	The density of a soil is its weight per cubic centimeter. It is sometimes expressed as wet weight or the total weight including water. It is more commonly the "dry weight", which is the weight of the soil particles alone, excluding the weight of the water.
liquid limit of soil	The liquid limit of soil is defined as the moisture content at which a soil passes from the plastic to the liquid state as defined by the liquid limit test.
plastic limit of soil	Plastic limit is again a property of cohesive soils and is defined as the moisture content at which a soil becomes too dry to be a plastic condition as defined by the plastic limit test.

For all new construction it is very important that in-situ moisture contents, Atterberg limits and grain size analysis of subgrade soil materials be determined to assess subgrade soil characteristics and to infer resilient modulus (MR) values.

2. Subgrade Strength Evaluation

The characteristic material property of subgrade soils used for pavement design is the resilient modulus (MR). The resilient modulus is defined as being a measure of the elastic property of soil recognizing selected non-linear characteristics. Methods for the determination of MR are described in AASHTO T294-92 test method. For many years, standard California Bearing Ratio (CBR) tests were used to measure the subgrade strength parameter as a design input.

For roadbed materials, the AASHTO Guide recommends that the resilient modulus be established based on laboratory testing of representative samples in stress and moisture conditions simulating the primary moisture seasons. Alternatively, the seasonal resilient modulus values may be determined based on correlations with soil properties.

For the design of new construction pavement structures, the subgrade resilient modulus is estimated using an existing representative roadway located near the new project, with similar subgrade soils and drainage conditions, as a prototype.

The prototype can be tested with the *FWD*[2] (Fig.3-15) and the deflection data analyzed in by computer program to determine the backcalculated subgrade modulus. This value can then be used as an approximation of the strength of the subgrade materials that will exist in the new subgrade.

3. Swelling Soil Potential

Excessively expansive soils such as highly plastic clays or bentonitic shales require special attention particularly when in close proximity to the surface of the road embankment. These materials contain what result in volume changes (swelling and shrinking) with changes in moisture content.

The need to control the intrusion of moisture into such soils is very important in order to mitigate swelling.

Fig.3-15 A FWD, towed by a truck

4. Frost Susceptibility

The climate in Heilongjiang province results in freezing of near surface subgrade soils for several months each year. The depth of frost penetration generally increases from the south to the north of the province.

Although some volumetric expansion occurs according to the freezing, a more significant issue is related to the spring melting period. If the structure has not been designed to account for weakened subgrade support, the thaw will release excess water which causes a loss of subgrade strength and potential damage to the roadway pavement structure.

The term frost heaves refers to the upward vertical displacement of a pavement surface as a direct result of the formation of ice lenses in a frost susceptible subgrade. True frost heave should have the following three factors: a frost susceptible soil, slowly depressed air temperatures, and a supply of water(Fig.3-16).

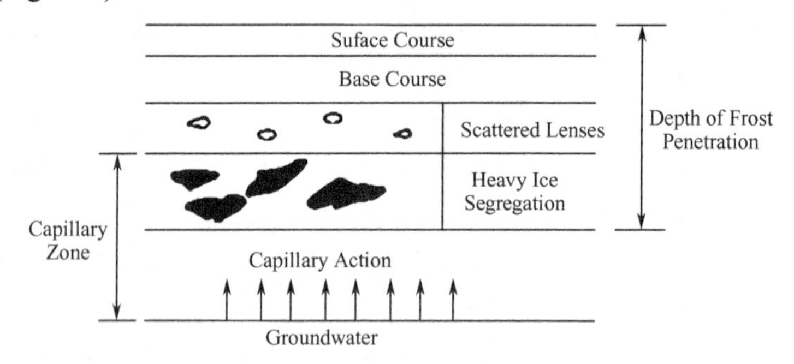

Fig.3-16 Formation of Ice Lenses

The removal of any one of the above three factors will reduce the potential for frost heaving and resulting surface distress.

Highway Subgrade Construction

1. Highway subgrade excavation

Subgrade excavation shall conclude excavating the roadbed, borrow pits, sidestep, borrow ditches and canals; broadening original subgrade; moving and dealing with the disposing of all surplus materials taken from the range of construction(Fig.3-17).

Subgrade excavation shall include all excavation, shaping and sloping which are necessary for the construction, preparation and completion of all subgrades. Shoulder slopes, road intersections and approaches, in close conformity to the alignment, levels and cross sections shown on the drawings.

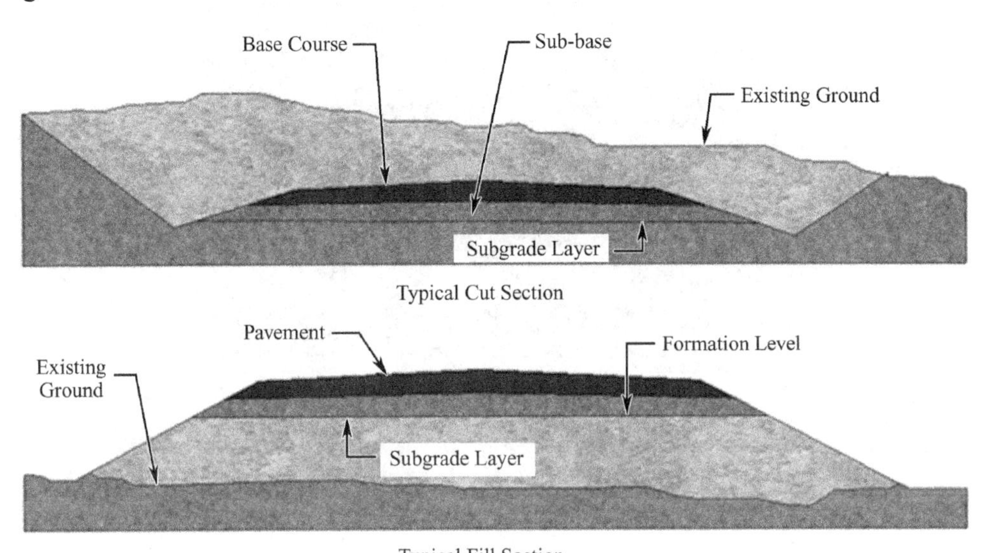

Fig.3-17　Typical Cut & Fill Section

2. Embankment filling

These projects conclude all subgrades utilization earth, rock filling, and borrow filling and relevant construction works (Fig.3-17).

Earth embankment must be filled in layers according to the design section and compacted by machines, its maximum loose thickness (< 30 cm) for a layer shall be determined in accordance with soil classifications, functions of compaction machines and rolling passes.

When roadway embankments are placed with permeable filling materials such as crushed stone, pebble, gravel and coarse sand, they may be not restricted by water content, but shall be filled and compacted in layers. When less permeable materials are used for placing embankment, the water content shall be under control to approach the optimal water content for compaction.

Rollers or other compaction devices increase soil density by expelling air from the voids in the soil and by rearranging or forcing the soil grains into more closely contact. Water aids as a lubricant up to the optimum moisture content. In porous soils, air is easily forced out, but in heavy or tight, cohesive soils more effort is required. Because of this, heavy cohesive materials must be placed in many layers if the air is expected to be expelled (Fig.3-18).

Compactors include tamping or sheep-foot rollers, pneumatic-tired rollers, steel-wheeled rollers (Fig.3-18), plate compactors, grid rollers and vibratory compactors. Sometimes two types of compactors, such as steel wheels and pneumatic tires are combined in a single unit.

Fig.3-18 Highway subgrade is compacted

3.3.2 Notes

1．土的工程分类请参照我国现行标准：《土的工程分类标准》（GB/T 50145—2007）。

2．FWD：落锤式弯沉仪（Falling Weight Deflectometer，简称 FWD）产生于 20 世纪 70 年代初，是一种脉冲动力弯沉仪，它模拟汽车载荷对路面施加瞬时冲击的作用，得到路面瞬时变形的情况。《公路路基路面现场测试规程》（JTG 3450-2019）中已将 FWD 列为弯沉检测设备。

3.3.3 New Words and Expressions

basement	【建】基底，底部
cut	挖方
borrow pit	取土坑
compact	压实
density	密度
moisture content	含水量
cross section	横断面
embankment	路堤
trench	沟渠
wearing course	磨耗层
wheel load	轮载荷
subsidence	下沉，沉降
grain size	粒径
cobble	卵石

gravel	砾石
silt	粉质土
clay	黏性土
mass	【物】质量；大量，众多
liquid limit	液限
plastic limit	塑限
ratio	比率
subtraction	减法
cubic centimeter	立方厘米
cohesive	黏性的，黏质的，有黏着力的
in-situ	原位，现场
in-situ moisture content	原位含水量
subgrade strength	路基强度
resilient modulus (MR)	回弹模量
non-linear	非线性
California Bearing Ratio (CBR)	加州承载比
prototype	原型
tow	拖拽
swelling soil	膨胀土
bentonitic	（含有）膨润土（皂土、斑脱土）的
shale	页岩
in close proximity to	与……靠得很近
volume	体积，容积
swelling	膨胀的，增大的
mitigate	减轻
susceptibility	敏感性
frost	霜冻
penetration	渗透
volumetric	测定体积的
thaw	解冻，融雪
frost heave	冻胀，冰冻膨胀
lens	透镜，镜头，透镜体
excavation	开挖
sidestep	侧向台阶
ditch	沟渠
canal	运河，沟渠
dispose	处置，处理

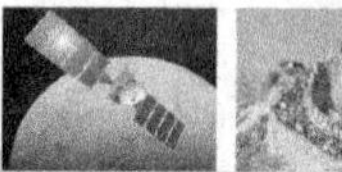

surplus	过剩的；多余的
shoulder slope	路肩横坡
road intersection	道路交叉口
approach	引道，引路
in close conformity　to	与……密切一致
drawing	图样，图纸
permeable	可渗透的
pebble	卵石
optimal	最佳的，最优的；最理想的
roller	滚筒，路碾，压路机
compaction device	压实设备
expel	排出
void[vɔɪd]	空的，空间
soil grain	土粒
lubricant [ˈlu:brɪkənt]	润滑剂，润滑油
optimum [ˈɒptɪməm]	最适宜的，最佳的
porous[ˈpɔ:rəs]	多空隙的，多孔的，能穿透的，能渗透的
compactor	【美】垃圾捣碎机，压土机，夯土机；夯具
tamping roller	羊足路碾，夯击式压路机
sheep-foot roller	羊足路碾，羊足压路机
pneumatic-tired roller	胶轮压路机
steel-wheeled roller	钢轮压路机
plate compactor	板式压路机
grid roller	网格压印路碾，网格压路机
vibratory compactor	振动压路机
pneumatic tire	充气轮胎

3.4　Pavement Engineering **路面工程**

扫一扫看本节参考译文

扫一扫看本节教学课件

3.4.1　Text

The main function of highway pavement is to provide a safe, stable and durable surface over which traffic may move for a period of time under the action of weather and a large number of vehicles. Highway pavements are divided into two main categories: rigid and flexible pavement. The wearing surface of a rigid pavement is usually constructed by Portland cement concrete and acts as a beam over any rough underlying supporting materials. The wearing surface of flexible pavement, on the other hand, is usually constructed by bituminous substances. Flexible pavement usually consists of two bituminous layers: a bituminous layer with granular material and a layer of a suitable mixture of coarse and fine materials. Traffic loads are transferred by the wearing surface to the underlying supporting materials through the interlocking of aggregates, the frictional effect of the granular materials, and the cohesion of the fine materials.

Flexible Pavement

Flexible pavements are constructed of bituminous and granular materials. The first asphalt road in the United States was constructed in 1870 at Newark, New Jersey. The first sheet-asphalt pavement which is a hot mixture of asphalt cement with clean, angular, graded sand and mineral filler was laid in 1876 on Pennsylvania Avenue in Washington DC with imported asphalt form Trinidad Lake. In 2001, there are about 2.5 million miles of paved roads in the United States, of which 94% are asphalt surfaced. China laid asphalt pavement in shanghai in 1920, and currently asphalt pavement (Fig.3-19) is widely used.

Fig.3-19　Asphalt surface

Conventional flexible pavements are layered systems with better materials on the top where the intensity of stress is high and inferior materials at the bottom where the intensity is low. *The components of a flexible pavement include the subgrade or prepared roadbed, subbase, base course and surface course*[1] (Fig.3-20). The performance of the pavement depends on the satisfactory performance of each component, which requires proper evaluation of the properties of each component separately.

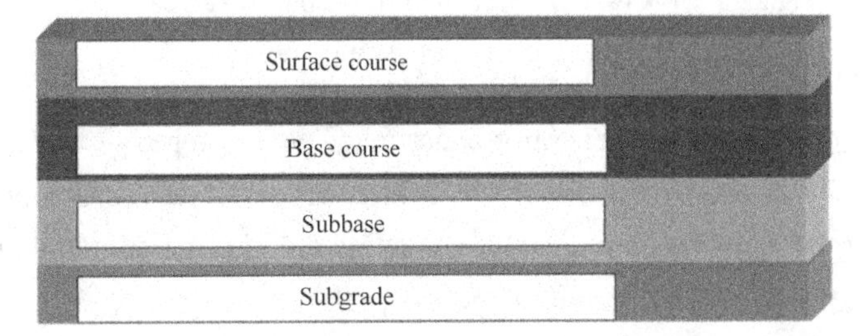

Fig.3-20　Typical cross section of conventional flexible pavement

Subgrade is consisted of the natural material locating along the horizontal alignment of the pavement and serves as the foundation of the pavement structure. The subgrade may also be a layer of selected borrow materials, well compacted to prescribed specifications. The top 152mm of

subgrade should be compacted to the desirable density near the optimum moisture content.

Subbase locates above the subgrade and consists of materials of a superior quality which is generally used for subgrade construction, the requirements of subbase materials are usually given in terms of the gradation, plastic characteristics and strength.

Base course lies immediately above the subbase. It is placed immediately above the subgrade if a subbase course is not used. This course usually consists of granular materials such as crushed stone (crushed or uncrushed). The specifications for base course materials usually include stricter requirements than those for subbase materials and their plasticity, gradation, and strength. In some cases, high-quality base course materials may also be treated with asphalt or Portland cement or lime to improve the stiffness characteristics of heavy-duty pavements.

Surface course is the upper course of the road pavement and is constructed immediately above the base course; the surface course in flexible pavement usually consists of a mixture of mineral aggregates and asphaltic materials. It should be capable of withstanding high tire pressures, resisting the abrasive forces due to traffic, providing a skid-resistant driving surface, and preventing the penetration of surface water into the underlying layers.

Full-depth asphalt pavements (Fig.3-21) are constructed by placing one or more layers of HMA directly onto the subgrade or improved roadbed. This concept was promoted by the Asphalt Institute in 1960 and is universally considered the most cost-effective and dependable type of asphalt pavement for heavy traffic. This type of construction is quite popular in areas where local materials are not available. It is more convenient to purchase only one material rather than several materials from different sources to minimize the administration and equipment costs.

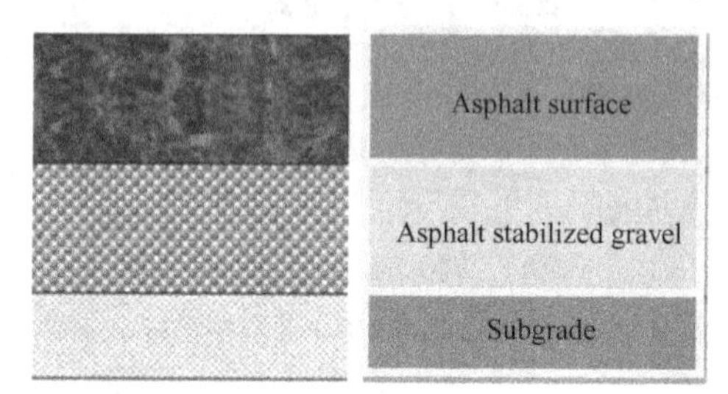

Fig.3-21 Typical cross section of full-depth asphalt pavement

Rigid pavement

Rigid highway pavements are generally constructed to carry heavy traffic loads. Originally, they have been used for residential and local roads. Properly designed and constructed rigid pavements have long service lives and usually are relatively cheaper to maintain than the flexible pavements.

The Portland cement concrete is often used for rigid pavements, which consists of Portland cement, coarse aggregate, fine aggregate, and water. Whether steel reinforcing rods are utilized or not depends on the type of pavement being constructed.

Rigid highway pavement is divided into three main types: plain concrete pavements, commonly reinforced concrete pavements, and continuously reinforced concrete pavements. The definition of each pavement type is related to the amount of reinforcement used:

Rigid pavement	plain concrete pavement	1. Plain concrete pavement has no temperature steel or dowels for load transfer. 2. Steel tie bars are often used to provide a hinge effect at longitudinal joints and to prevent the opening of these joints. 3. Plain concrete pavements are mainly used on low-volume highways or when cement-stabilized soils are used as subbase. 4. Joints are placed for relatively shorter distances (10 to 20 ft) than the other types of concrete pavements to reduce the amount of cracking.
	commonly reinforced concrete pavement	1. Simply reinforced concrete pavements have dowels for the transfer of traffic loads across joints, with these joints spaced for larger distances, ranging from 30 to 100 ft. 2. Temperature steel is used throughout the slab, with the amount dependent on the length of the slab. 3. Tie bars are also commonly used in longitudinal joints.
	continuously reinforced concrete pavement	1. Continuously reinforced concrete pavements have no transverse joints, except for construction joints or expansion joints when they are necessary at specific positions, such as at bridges. 2. These pavements have a relatively high percentage of steel, with the minimum usually at 0.6 percent of the cross section of the slab. 3. They also contain tie bars across the longitudinal joints.

3.4.2 Notes

1．常规柔性路面的标准结构层从顶部开始，路面由沥青层、上面层、黏结层、中面层、透层、基层、底基层、压实的路基和自然路基构成。选用不同的结构类型时以必要性或经济性为基础，因此有些结构层可能会被略去。

2．在 20 世纪五六十年代我国道路的沥青层厚度差异非常大，主要以 15 cm 的沥青层为主。在 2000 年以后，沥青层有增厚趋势，4 cm+6 cm+8 cm 成为主流结构。主要考虑两点，原 4 cm+5 cm+6 cm 对于中下面层沥青混合料的施工厚度与混合料的公称粒径相比偏薄，不利于碾压；另外受欧日美地区厚沥青层的影响，有意识地增加沥青层厚度到 17 cm、18 cm。自 2005 年至今，进入厚沥青层时代。交通部 2005 年 11 月发布“关于防治高速公路沥青路面早期损坏的指导意见”的通知，提出“完善结构和厚度设计”，“鼓励各地加强柔性基层试验研究，在试验路段铺筑成功的基础上加以推广”等，使得全国对沥青路面结构形式和沥青面层的厚度有了新的认识。大部分省的路面沥青层厚度增加到 18 cm，部分省的路面沥青层厚度达到 20 cm 及以上，山东省的某些道路甚至达到 32 cm 及以上；但西部省份交通量不大的路段仍有采用 9 cm、12 cm 的。

3.4.3 New Words and Expressions

weather	风化
wearing surface	磨耗面，磨损面
Portland cement concrete	普通硅酸盐混凝土

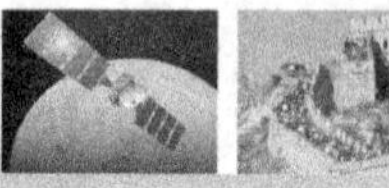

underlying	基础的；表面下的，下层的
granular	粒状的，颗粒的
coarse	粗的
fine	细的
traffic load	交通载荷
transfer	使转移
interlock	互锁，嵌锁
frictional ['frɪkʃənəl]	摩擦的，摩擦力的
cohesion [kəʊˈhiːʒn]	凝聚，内聚；（各部的）结合；【力】内聚力
sheet-asphalt pavement	层状沥青路面
prescribe [prɪˈskraɪb]	规定，指定
requirement	要求，规范
gradation	级配
plastic	塑性的
base	基层
plasticity	塑性
high-quality	高质量
stiffness	坚硬
heavy-duty	重型的，重载的
mineral aggregate	矿料
asphaltic	柏油的
tire pressure	轮胎气压
abrasive force	摩擦力
skid-resistant	防滑的
full-depth asphalt pavement	全厚式沥青路面
cost-effective	有成本效益的，划算的；合算的
dependable	可靠的
heavy traffic	大流量交通
residential	住宅的
service life	使用寿命
maintain	维护，保养
coarse aggregate	粗集料
fine aggregate	细集料
steel reinforcing rod	钢筋条
utilize	使用
plain concrete	素混凝土
dowel	木钉，销子

hinge	铰链，合页；关键，转折点
joint	关节；接合处
cracking	破裂，裂缝
longitudinal	经度的；纵向的
transverse	横向的
expansion joint	伸缩接头，伸缩缝

扫一扫看本节参考译文

3.5 Highway Interchange 公路互通式立体交叉

扫一扫看本节教学课件

3.5.1 Text

An *intersection*[1] is the crossing where two or more roads join or cross, including the road and roadside facilities for traffic movements (Fig.3-22). The primary step of controlling traffic volume includes discharging mode design, signal design and channelization design of signalized intersection. *Channelization*[2] of an intersection at grade is to direct traffic into defined paths by the use of islands or traffic markings.

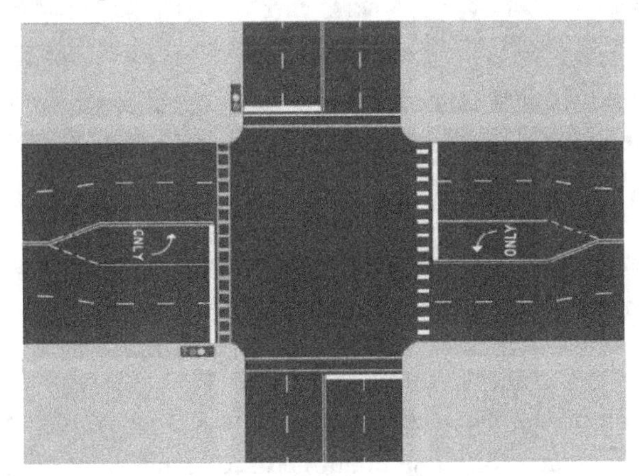

Fig.3-22　Intersection

A traffic interchange is a combination of ramps, grade separations and other facilities at the junction of two or more highways for the purpose of reducing or eliminating traffic jams to improve safety and increase traffic capacity (Fig.3-23). Crossing conflicts are eliminated by grade separation design. Turning conflicts are either eliminated or minimized according to the type of interchange design. In contrast, at-grade intersections must handle a variety of conflicts among vehicles, pedestrians and bicycles.

All connections to freeways are done by traffic interchanges. An interchange or separation may be warranted as a part of an expressway, to improve safety or eliminate a bottleneck, or where topography does not allow the construction of an intersection.

The minimum interchange intervals shall be 1.5 km in urban areas, 3.0 km in rural areas and 3.0 km between freeway-to-freeway interchanges and local street interchanges.

Fig.3-23 A multi-level stack interchange in Shanghai, China.

To improve operations of interchanges in closely range, the use of auxiliary lanes, grade separated ramps and ramp metering may be warranted.

Terminology

1. A ***freeway junction*** or ***highway interchange*** (in the US) or ***motorway junction*** (in the UK) is a kind of road junction, connecting one controlled-access highway (freeway or motorway) to other roads or to motorway service area.

2. A ***highway ramp*** (as in ***exit ramp / off-ramp*** and ***entrance ramp / on-ramp***) or ***slip road*** is a short section of road which allows vehicles to enter or exit a controlled-access highway (freeway or motorway) (Fig.3-24).

3. A ***directional ramp*** makes a left turn to exit from the left side of the roadway (a left exit).

4. A ***non-directional ramp*** goes in a direction opposite to the desired direction of traveling and makes a left turn to exit from the right side of the roadway. Many loop ramps are non-directional.

5. A ***semi-directional ramp*** exits a road in a direction opposite from the desired direction of traveling, but then turns toward the desired direction of traveling. Many flyover ramps (as in a stack) are semi-directional.

6. A ***U-turn ramp*** leaves the road in one driving direction, turns over or under it and rejoins in the opposite direction.

Fig.3-24 Off-ramp accessed from collector/distributor lanes along Highway 401 in Toronto, Ontario, Canada.

Interchange Types

The selection of an interchange type and its design is influenced by many factors including the following: the speed, volume and composition of traffic in service, the number of intersecting roads, the standard and arrangement of the local street system including traffic control devices, topography, right of way controls, local planning, proximity of adjacent interchanges, community impact, and cost.

Basic freeway-to-freeway interchange design configurations are illustrated in the following table. Many combinations and variations may be formed from these basic interchange types.

Four-way interchanges	Cloverleaf interchange	
	Stack interchange	
	Turbine interchange	

续表

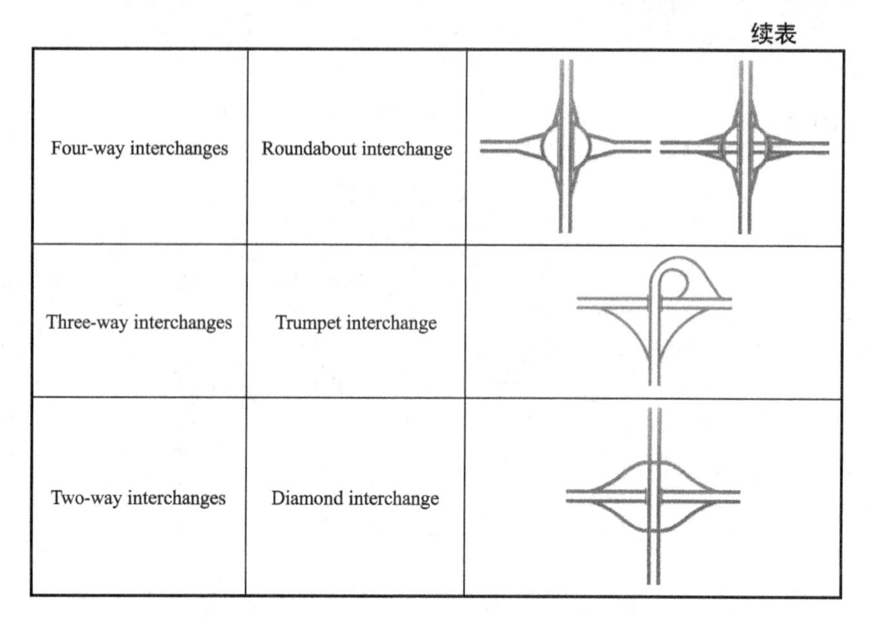

Four-way interchanges	Roundabout interchange	
Three-way interchanges	Trumpet interchange	
Two-way interchanges	Diamond interchange	

1. A ***cloverleaf interchange*** is typically a two-level, four-way interchange where all turns across opposing traffic are handled by non-directional loop ramps (Fig.3-25). Assuming right-handed traffic to go left, vehicles first cross over or under the target route, then bear right onto a sharply curved ramp that turns roughly 270 degrees, merging onto the target route from the right, and crossing the route just departed. These loop ramps produce the namesake cloverleaf shape.

Fig.3-25　A typical cloverleaf interchange with collector roads in Michigan

2. A ***stack interchange*** is a four-way interchange whereby a semi-directional left turn and a directional right turn are both available (Fig.3-26). Usually access to both turns is provided

simultaneously by a single off-ramp. In order to cross over incoming traffic and go left, vehicles from right-handed driving first exit onto an off-ramp from the rightmost lane. After demerging from right-turning traffic, they complete their left turn by crossing both highways on a flyover ramp or underpass. The penultimate step is a merge with the right-turn on-ramp traffic from the opposite quadrant of the interchange. Finally an on-ramp merges both streams of incoming traffic into the left-bound highway.

Fig.3-26　Yan'an East Road Interchange, seen from a pedestrian's perspective

3. The ***turbine/whirlpool interchange*** which is a derivative of stack interchange requires fewer levels (usually two or three) while retaining semi-directional ramps throughout, and has its left-turning ramps sweep around the center of the interchange in a spiral pattern (Fig.3-27).

Fig.3-27　The Interchange in Chicago, a notable turbine interchange

4. ***Roundabout interchange***: The ramps of the interchanging highways meet at a roundabout or rotary on a separated level above, below, or in the middle of the two highways (Fig.3-28).

Fig.3-28　Complex roundabout interchange in the Netherlands

5. ***Trumpet interchanges*** are named as such due to their resemblance to trumpets. These interchanges are very common on toll roads, as they concentrate all entering and exiting traffic into a single stretch of roadway, where *toll plazas*[3] can be installed once at one side to handle all traffic from both directions, especially on ticket-based tollways (Fig.3-29).

Fig.3-29　A trumpet interchange on the Sir John A. Macdonald Parkway in Ontario

6. A ***diamond interchange*** is an interchange involving four ramps where they enter and leave the freeway at a small angle and meet the non-freeway at almost right angles (Fig.3-30). These ramps at the non-freeway can be controlled through stop signs, traffic signals, or turn ramps.

Fig.3-30　Diamond interchange on Interstate 71 in West Lancaster

3.5.2　Notes

1. 单词区分

intersection	An intersection is **an at-grade junction** where two or more roads meet or cross. Intersections may be classified by number of road segments, traffic controls, and/or lane design. 交叉路口是指两条或两条以上道路交会或交叉的**平面交叉路口**。交叉路口可按路段数、交通管制和/或车道设计分类。
interchange	An interchange is a road junction that typically uses **grade separation**, and one or more **ramps**, to permit traffic on at least one highway to pass through the junction without directly crossing any other traffic stream. **互通式立体交叉道**是路线交叉中的一种类型，交叉道路之间立体交叉并以一条或多条匝道相互连通以实现交通转换，允许至少一条公路上的车辆以不直接穿过任何其他交通流的方式通过交叉路口。

2. 渠化（Channelization）是通过导流岛与路面标线相结合的方式，以分隔或控制冲突的车流，使之进入规定的路线，从而满足平面交叉的基本要求。

3. toll plaza: a row of tollbooths across a road（道路上的）收费站，收费区，收费广场。

3.5.3　New Words and Expressions

roadside	路边，路旁
facility	设施，设备
signal	信号
channelization	渠化
ramp	匝道
eliminate	消除
traffic jam	交通堵塞

traffic capacity	交通（容）量
minimize	最小化
at grade	【美】在同一水平面上
vehicle	车辆，交通工具
pedestrian	行人
freeway	高速公路
warrant	保证
expressway	高速公路
bottleneck	瓶颈
ramp metering	匝道车流调节
multi-level	多层的
freeway junction	高速公路立体交叉路口
highway interchange	公路立体交叉道
motorway junction	高速公路交叉路口
controlled access highway	控制驶入的公路
motorway service area	高速公路服务区
highway ramp	公路匝道
exit ramp	出口匝道
off-ramp	出口匝道
entrance ramp	入口匝道
on-ramp	入口匝道
slip road	高速公路会交点
directional ramp	直接式匝道
non-directional ramp	非直接式匝道
semi-directional ramp	半直接式匝道
U-turn ramp	回转匝道
flyover	立交桥，高架公路
collector lane	辅道
distributor lane	辅道
configuration	构造，外形，布局
four-way interchange	四方向立体交叉道
cloverleaf interchange	苜蓿叶形立体交叉道
stack interchange	环状形立体交叉道
turbine interchange	漩涡形立体交叉道
roundabout interchange	环岛形立体交叉道
three-way interchange	三方向立体交叉道
trumpet interchange	喇叭形立体交叉道

two-way interchange	双方向立体交叉道
diamond interchange	钻石形立体交叉道
merge	（使）混合；相融；融入
loop ramp	环形匝道
namesake	同姓名的人；同名的事物
simultaneously [ˌsɪməl'teɪnɪəslɪ]	同时的，一致的
rightmost	最右边
demerge [ˌdiː'mɜːdʒ]	拆分，分离
right-turning	右转
flyover ramp	立交桥匝道
underpass	地下通道
penultimate [pen'ʌltɪmət]	倒数第二（的）
quadrant	象限
left-bound highway	左行公路
derivative [dɪ'rɪvətɪv]	衍生物，派生物
sweep	蜿蜒
spiral['spaɪrəl]	螺旋形的；盘旋的
rotary['rəʊtəri]	旋转的；环行交叉路
roundabout	绕道；环形交通枢纽
trumpet ['trʌmpɪt]	喇叭
toll road	收费公路
toll plaza	（道路上的）收费站，收费区，收费广场
tollway	【美】收费公路
non-freeway	非高速公路
right angle	直角

3.6 Highway Maintenance& Management 道路养护与管理

3.6.1 Text

扫一扫看本节参考译文

扫一扫看本节教学课件

Surface deterioration

Like all structures, highway systems fail due to fatigue over time. Deterioration appears as a result of accumulated damage from the traffic loads, environmental effects such as frost heaves, thermal cracking and oxidation.

Other failure modes include aging and surface abrasion. If no maintenance is done promptly on the wearing course, potholes will form. The freeze-thaw cycle in cold climates will dramatically accelerate pavement deterioration, once water can penetrate the surface. *The early damage of asphalt pavement is divided into two aspects: functional and structural damage*[1]. Common types of

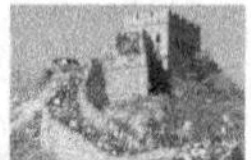

distresses in highway pavement are listed below:

1. ***Rutting*** is defined as permanent deformation resulting from the load of vehicle wheels on the roadway, unstable HMA, densification of HMA and deep settlement in the subgrade (Fig.3-31).

Fig.3-31 14 cm rut on a main road

2. ***Bleeding*** is the upward movement of the asphalt binder and results in the formation of a film of asphalt on the surface of road. Bleeding appears when the HMA mix is added excess asphalt binder which is forced onto the surface by the traffic loads in hot weather (Fig.3-32).

Fig.3-32 Bleeding phenomenon of asphalt pavement

3. ***Frost heave*** is an upward swelling of soil during freezing conditions caused by an increasing of ice as it grows towards the surface, upwards from the depth in the soil where freezing temperatures are penetrated into the soil (Fig.3-33, Fig.3-34).

Fig.3-33 Anatomy of a frost heave during spring thaw

Fig.3-34 Asphalt damaged by frost heaves

4. ***Fatigue cracking*** is an asphalt pavement distress most often instigated by failure of the surface due to traffic loading. And, fatigue cracking can be greatly influenced by environmental and other effects while traffic loading remains the direct cause (Fig.3-35).

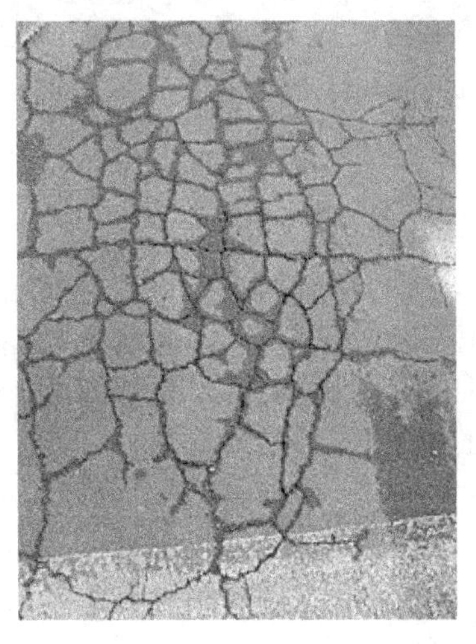

Fig.3-35 Moderate to severe Fatigue cracking

5. A ***pothole*** is a structural failure in a highway surface, usually on asphalt pavement, due to water in the underlying soil structure and traffic passing over the influenced area. Water first weakens the underlying soil, traffic then fatigues and breaks the poorly supported asphalt surface in the influenced area. Continued traffic load ejects both asphalt and the underlying soil material to create a hole in the pavement (Fig.3-36, Fig.3-37).

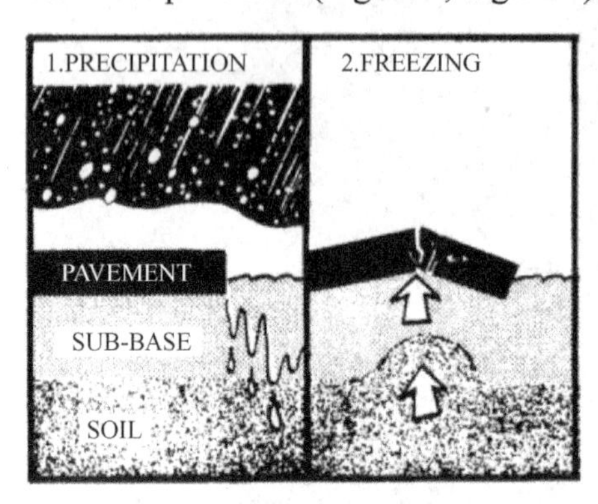

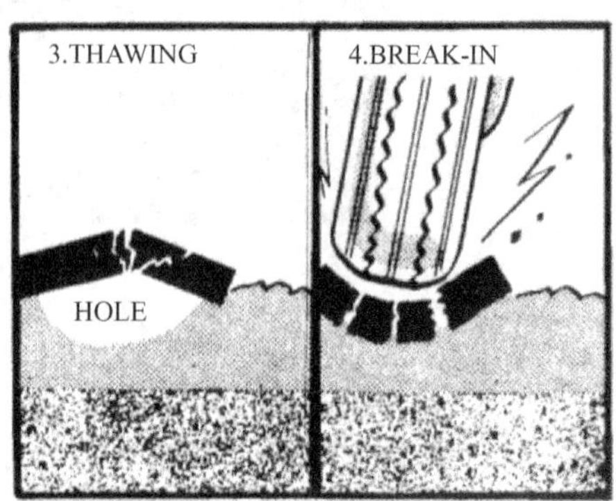

Fig.3-36 Factors leading to pothole failure by fatigue in areas subject to freezing and thawing are: 1. Precipitation adds moisture to supporting soil structure. 2. Frost heaving can damage pavement. 3. Thawing can weaken soil structure. 4. Traffic can break the pavement.

Fig.3-37 A pothole reappearing on a newly patched roadway, also showing the transition between crocodile cracking and the pothole, with water dried up

Highway Maintenance

Pavements are designed for an expected service life or design life. In some parts of the UK, the standard design life is 40 years for new bitumen and concrete pavement. Highway maintenance is the task of preserving, repairing and restoring a system of roadways and its elements, to its designed or accepted configuration. Roadway elements are as follows: carriageway surfaces, shoulders, roadsides, drainage facilities, bridges, tunnels, signs, markings and lighting fixtures.

Highway maintenance program is developed to carry out the above tasks and to suppress the detrimental effects of weather, deterioration, traffic wear, damage and *vandalism*[2]. Deterioration includes the effects of aging, material failures, and design and construction faults. The preservation

and repair of buildings, stockpiles and equipment essential to performing the highway maintenance task are also parts of highway maintenance program.

A routine maintenance program is carried out as frequently as required during each year on all elements of the highway, in order to ensure serviceability at all times and in all weathers. The main operations included are:

1. The cleansing of carriageways, ditches, drains, signs and signals, safety barriers, as well as grass trimming and tree pruning;
2.The repair of minor damage to carriageways, slopes, culverts, signals and signposts, safety barriers, lighting facilities and buildings, as well as any urgent interventions required to restore disrupted traffic movement(removal of debris from the carriageway);
3. The replacement of ancillary furniture and equipment that has been damaged, e.g. signing, barriers, road markings, drainage tubes, small channels, planted areas, lighting facilities;
4. In winter, maintenance operations are intended to retain serviceability in poor weather conditions (for example, clearance of snow and ice) for prevention and cure.

A ***periodic maintenance program*** covers all long-term maintenance operations required within the service life of the highway. These activities, which may be required only at intervals of several years, can be divided into two main groups as follows:

1. The renewal or renovation of wearing surfaces of carriageways that become worn or deformed by travel loads;
2. The restoration of road markings, culverts and ancillary items and the repainting of metal bridges etc.

Extraordinary maintenance activities aim to refurbish roads to their original condition when they have been severely deteriorated. Generally, they involve following items：

1. The strengthening or reconstruction of a pavement structure that has deteriorated severely;
2. Main actions to protect roads against external agents (for example, actions involving slop stabilization and falls of rocks, retaining walls) and protection against flooding and avalanches.

Maintenance for the older concrete pavements that develop faults includes the technique called dowel bar retrofit. This involves cutting slots in the pavement at each joint, placing dowel bars in the slots, and then filling them with concrete patching material. This method can extend the life of the concrete pavement for another 15 years.

3.6.2 Notes

1．沥青路面早期损坏现象分为功能性和结构性破坏两个方面：

<table>
<tr><td rowspan="5">功能性损坏</td><td colspan="2">泛油，表面构造衰减、松散剥落</td></tr>
<tr><td colspan="2">平整度衰减</td></tr>
<tr><td colspan="2">车辙</td></tr>
<tr><td rowspan="2">表面开裂</td><td>温度收缩</td></tr>
<tr><td>反射裂缝</td></tr>
<tr><td rowspan="4">结构性损坏</td><td colspan="2">路基引起的路面结构发生不均匀沉降</td></tr>
<tr><td colspan="2">基层损坏引起的道路表面出现坑槽</td></tr>
<tr><td colspan="2">沥青面层剪切破坏</td></tr>
<tr><td colspan="2">桥面铺装损坏</td></tr>
</table>

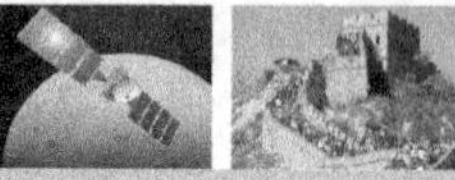

2. vandalism: the crime of destroying or damaging something, especially public property, deliberately and for no good reason 故意破坏公共财物罪；恣意毁坏他人财产罪。

3.6.3 New Words and Expressions

maintenance	维持，保持；保养，保管；维护；维修
deterioration [dɪˌtɪərɪə'reɪʃn]	恶化；变坏；退化
fatigue [fə'ti:g]	疲劳
thermal cracking	加热分裂（法），热裂化，热裂解，热破裂
oxidation[ˌɒksɪ'deɪʃn]	氧化
aging	老化；老龄化；（酒等的）陈化；熟化
surface abrasion	路面磨损
promptly ['prɒmptli]	迅速地；立即地
pothole ['pɒthəʊl]	（路面的）坑槽
freeze	冻结；严寒时期
thaw [θɔ:]	解冻，融解，回暖
penetrate	穿透，刺入；渗入
distress	悲痛；危难，不幸
rutting	车辙
densification [densɪfɪ'keɪʃən]	增浓作用，稠化（作用）；致密；捣实；夯实
settlement	沉淀，沉降
bleeding	泛油
upward	向上的
asphalt binder	沥青结合料
formation	形成；构成，结构
film	薄层；薄膜
excess [ɪk'ses]	超过（的）；超额量（的）
phenomenon [fə'nɒmɪnən]	现象
swell [swel]	增强；肿胀；膨胀
coalesce [ˌkəʊə'les]	联合，合并
anatomy [ə'nætəmi]	解剖，分解，分析；（详细的）剖析
instigate ['ɪnstɪgeɪt]	教唆；煽动；激起
moderate ['mɒdərət]	有节制的；稳健的，温和的；适度的，中等的
severe	严重的
eject[i'dʒekt]	喷出；驱逐；强制离开
moisture	水分，湿气
patch	补丁，修补
transition	过渡，转变，变迁
crocodile cracking	龟裂
preserve	保护；保持，保存
repair	修理；纠正；恢复；弥补
restore	归还；交还；使恢复；修复

element	元素
carriageway	行车道
shoulder	路肩
drainage[ˈdreɪnɪdʒ]	排水系统
tunnel	隧道
marking	标记，记号
lighting fixture	照明器材
suppress [səˈpres]	镇压，压制；禁止（发表）；阻止……的生长（或发展）
detrimental [ˌdetrɪˈmentl]	有害的（人或物）；不利的（人或物）
traffic wear	交通磨损
vandalism [ˈvændəlɪzəm]	故意破坏公共财物罪；恣意毁坏他人财产罪
preservation [ˌprezəˈveɪʃn]	保存，保留；保护；防腐；维护，保持
stockpile [ˈstɒkpaɪl]	n.（原料，食品等的）储备，备用物资；v.大量储备
routine maintenance	日常养护
serviceability [sɜ:vɪsə'bɪlɪtɪ]	有用性，适用性；可维护性
cleanse [klenz]	净化；使……清洁；清洗
safety barrier	安全屏障
trim	修剪；整理
prune [pru:n]	修剪（树木等）
culvert[ˈkʌlvət]	涵洞
signpost	指示牌，标志杆；路标；
urgent	急迫的；催促的
intervention	介入，干涉，干预；调解，排解
disrupt	v.使中断；adj.中断的
debris [ˈdebri:]	碎片，残骸
ancillary[ænˈsɪləri]	辅助的；补充的；附属的；附加的
road marking	路标
drainage tube	排水管
channel	沟渠
periodic maintenance	定期养护
long-term	长期
renewal[rɪˈnju:əl]	重建，重生；更新，革新
renovation [ˌrenəˈveɪʃn]	翻新，修复，整修
deform	使变形
refurbish[ˌri:ˈfɜ:bɪʃ]	刷新；使重新干净
strengthen	加强，巩固
reconstruction	重建；再现；重建物；复原物
external [ɪkˈstɜ:nl]	外面（的），外部（的）
stabilization [ˌsteɪbəlaɪˈzeɪʃn]	稳定性；稳定化
retaining wall	挡土墙
avalanche [ˈævəlɑ:nʃ]	雪崩

dowel bar	（混凝土路面）销钉，接缝条；传力杆
retrofit [ˈretrəʊfɪt]	翻新，改型
slot	窄缝
joint [dʒɔɪnt]	关节；接合处

知识分布网络

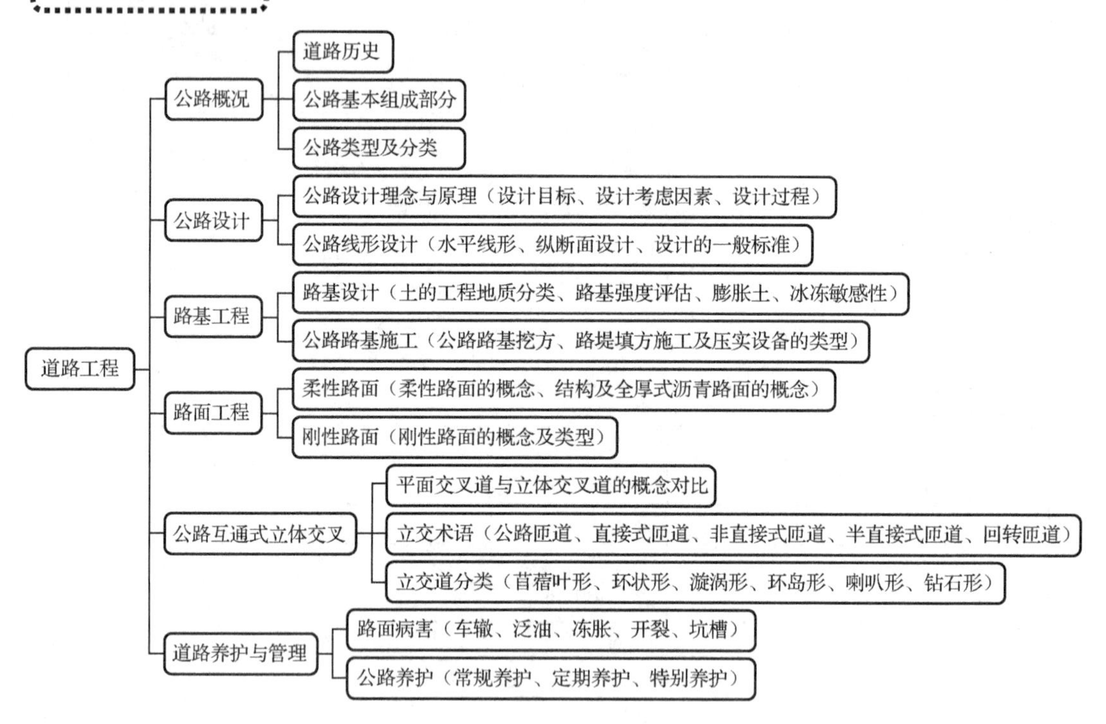

知识梳理与总结

本章内容是该教材的重点部分，课文均取自近年英文原版书刊及网络，基本上包含了道路工程的常用专业英语词汇。取材难度适中，既注重对学生英语语言的训练，又重视道路工程相关专业领域最新知识的传递。在每篇文章后都附有专业英语词汇表和注释，希望学生能够明确学习目标，依据教材、词典、多媒体课件、视频等学习手段和工具掌握道路工程相关专业的英语词汇，并且可以阅读简单的英语文献资料。本章必须掌握的知识要点有：

1. 在公路概述部分，需要掌握公路的发展历史、公路的基本组成部分和公路类型等知识点及专业英语词汇；

2. 公路设计是道路勘测设计的主要内容，本教材从公路设计理念与原理和公路线形设计两个方面进行了阐述，要求学生掌握公路的设计目标、设计考虑因素、设计过程、水平线形、纵断面设计、设计的一般标准等内容及专业英语词汇；

3. 路基工程包括路基设计和公路路基施工两方面，分别从土的工程地质分类、路基强度评估、膨胀土、冰冻敏感性及公路路基挖方、路堤填方施工及压实设备的类型等角度进

行了介绍，希望学生通过学习可以掌握相关概念、施工过程、设备等的英语表达，并且能够理解及阅读相关文献、文件和资料；

4. 路面的主要作用是提供一个安全、稳定、耐久的行车环境，主要由面层、基层和底基层组成，可以使用沥青混凝土或水泥混凝土，通过对本章的学习，学生需要掌握柔性路面的概念、结构，全厚式沥青路面的概念及刚性路面的概念、类型；

5. 公路立体交叉是公路的重要组成部分之一，一座公路立体交叉是由两条或两条以上公路相交处的匝道与立交构成的，以达到减少或消除冲突点的目的，来提高安全性、增加交通通行能力。本章讲述了平面交叉道与立体交叉道的概念对比、公路匝道、直接式匝道、非直接式匝道、半直接式匝道、回转匝道等立交术语及立交道分类，希望学生可以掌握相关专业英语词汇并阅读理解相关专业英语文献资料；

6. 高速公路沥青路面存在损坏情况，高速公路的发展较快，使得养护维修成为公路建设的重要课题和内容。通过对本章的学习，希望学生可以掌握车辙、泛油、冻胀、开裂、坑槽等路面损坏及公路的常规养护、定期养护、特别养护等的知识点，并掌握一定的英语词汇量。

思考与练习题 3

扫一扫看第 3 章习题答案

Task 1: Match the words given in Column A with the meaning given in Column B.

A	B
1. longitude	a. the ability to last for a long time without breaking or getting weaker
2. fatigue	b. to provide a safe, stable and durable surface
3. linearity	c. a combination of ramps, grade separations and other facilities at the junction of two or more highways
4. collector lane	d. a junction where roads meet or cross at grade
5. durability	e. concrete pavement
6. central median	f. asphalt pavement
7. vertical curve	g. steel-wheeled roller
8. subgrade	h. the distance of a place east or west of the Greenwich meridian , measured in degrees
9. pavement	i. property which is represented by a straight line
10. intersection	j. subgrade median
11. interchange	k. the upper layer in roadway construction
12. stability	l. weakness which is caused by repeated stress
13. rigid pavement	m. main road
14. entrance ramp	n. the native material underneath a constructed road, pavement or railway track. It is also called formation level
15. flexible pavement	o. auxiliary road
16. off-ramp	p. a road which cars use to drive off a major road
17. wearing course	q. a usual disease of asphalt pavement

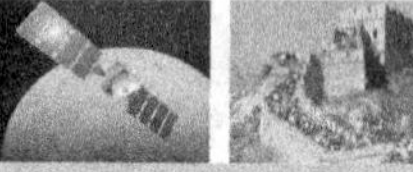

18. arterial road　　　　r. a road which cars use to drive onto a major road
19. compactor　　　　　s. the quality or state of being steady
20. bleeding　　　　　　t. a smooth transition curve between two tangent grades so as to provide a safe, comfortable sight distance

Task 2: Choose one correct answer from A, B, C and D.

1. What are the characteristics of Roman roads?______
(A) linearity and comfortable
(B) durability
(C) linearity and durability
(D) comfortable and durability

2. Why dose Roman road have very strong strength?______
(A) because of its pavement structure
(B) because of its pavement materials
(C) because of its pavement thickness
(D) all of the above

3. The subgrade soil is often referred to as______.
(A) the foundation or road bed soil
(B) parement materials
(C) UP-PCR materials
(D) road bed embankment

4. The prototype can be tested with the FWD and the deflection data analyzed with computer program to_______.
(A) determine the seasonal resilient modulus
(B) measure the elastic property of a soil
(C) determine the back-calculated subgrade modulus
(D) determine the subgrade modulus

5. Utilization of swelling materials in only lower portions of the embankment is often undertaken in order to______.
(A) result in volume changes(swelling and shrinking)
(B) make environment and drainage effects
(C) minimize these effects
(D) effect moisture contents

6. The need to control the intrusion of moisture into such soils is of major importance in order to______.
(A) result in swelling
(B) increase swelling
(C) reduce swelling
(D) mitigate swelling

7. Many factors______, affect the design of tangent and curve sections.

(A) including terrain conditions, physical features

(B) including terrain conditions, physical features and right-of-way considerations

(C) including terrain conditions

(D) including right-of-way considerations

8. An important element in ensuring driver safety and maintaining a roadway's operational efficiency is______the length of roadway ahead visible to the driver.

(A) providing adequate sight distance

(B) increasing safe distance

(C) providing space

(D) providing elevation of space

9. Turning conflicts arc either eliminated or minimized,______.

(A) depending upon the type of interchange design

(B) because of the traffic capacity

(C) in term of a unique characteristic of intersections

(D) depending upon the method of interchange design

10. Consideration should be given to providing_______in advance of the loop off-ramps to provide for vehicle deceleration.

(A) an lane

(B) an main lane

(C) one way lane

(D) an auxiliary lane

Task 3: Translate the following paragraphs into Chinese.

1. Rutting is defined as permanent deformation resulting from the load of vehicle wheels on the roadway, unstable HMA, densification of HMA and deep settlement in the subgrade.

2. An intersection is the crossing where two or more roads join or cross, including the road and roadside facilities for traffic movements. The primary step of controlling traffic volume includes discharging mode design, signal design and channelization design of signalized intersection. Channelization of an intersection at grade is to directed traffic into defined paths by the use of islands or traffic markings.

3. Highway pavements are divided into two main categories: rigid and flexible pavement. The wearing surface of a rigid pavement is usually constructed by Portland cement concrete and acts as a beam over any rough underlying supporting materials. The wearing surface of flexible pavement, on the other hand, is usually constructed by bituminous substances. Flexible pavement usually consists of two bituminous layers: a bituminous layer with granular material and a layer of a suitable mixture of coarse and fine materials. Traffic loads are transferred by the wearing surface to the underlying supporting materials through the interlocking of aggregates, the frictional effect of the granular materials, and the cohesion of the fine materials.

4. The characteristic material property of subgrade soils used for pavement design is the resilient modulus (MR). The resilient modulus is defined as being a measure of the elastic property of soil recognizing selected non-linear characteristics.

5. Highway Subgrade (or basement soil) is defined as the supporting structure of the layers of pavement. In cut section, the subgrade is the original soil lying below the layers designated as base and subbase. In fill sections, the subgrade is constructed over the nation ground with imported material from nearby roadway cuts or from borrow pit and compacted to a specified density and moisture content.

Task 4: Oral exercise. Please carry on the following dialogue with your partner about inspection of safety.

A: Firstly, I'd like to know the organization for safety works.

B: From the beginning of the project, we have appointed a qualified and experienced staff named Safety Officer who has sufficient time, authority and responsibility to ensure the safety program. The safety officer with his three assistants has an office in project management. There are totally ten full-time persons working for safety matters. Each of them is responsible for his/her own working section with around 150 to 200 workers.

A: That is good. As you know, setting up safety organization is important as well as giving people on site safety drill. Would you please give me a brief introduction to your complete training program of safety?

B: Since the setting up of the site, our Safety Training Program is carried out on the basis of the Standard JGJ59-2011 of the Ministry of Housing and Urban-Rural Development.

A: What are the specific contents of the safety training class?

B: The contents consist of the following items:jobsite safety policy, employee responsibility under Chinese Labour Law, company's safety regulations, electrical safety,construction accidents and reporting, scaffolding, trenching and excavation, crane safety, respiratory protection, personal safety equipment, fire protection and prevention, toxic substances, firs aid and emergency aid procedure, and so on.

A: Good. Now we can go to the work area for a safety inspection to see if the safety regulations are strictly implemented or not. Let's start with the helmet, the personal protective equipment. It's regret that not everyone wears the helmet and protective footwear.

B: Because the weather in this area is very hot, local people feel uncomfortable with shoes and helmet which are useful in protection of their health and safety.

A: You must take severe measures to ensure that everybody wears safety shoes and helmet. Besides, staff working in dusting areas must wear the respirators.

B: OK. I think our scaffolding is very satisfying.

A: No. I am not satisfied with the working condition. Some working tables above the height of 2 meters from ground are without ladders. And in a few places there are too many materials piled on the scaffolding.

B: It is really a dangerous operation, we must rectify it at once. An instruction will be given to the working section.

【注释】

1. 对话背景：在国际工程中常汇集不同国家、地区的技术人员，雇佣大量当地的劳务人员，他们的思想观念和工作方式不同，且存在语言障碍，在施工中出现事故的可能性很大。强调施工安全并不会影响工程的进度。对于危险施工处，要标有使用当地语言书写的警告标志牌（warning signs）。在进场道路和施工现场内的道路危险路段，也应设有限速标志牌（speed limit sign）。标志牌示例如图 3-38 所示。

Fig.3-38　"Road works ahead" sign, typically used in construction site

2. New words and expressions

sufficient　足够的，充足的

content　内容，目录；容量，含量

regulation　管理；规章，规则，章程

scaffold [ˈskæfəʊld]　脚手架

trench [trentʃ]　深沟，地沟

excavation [ˌekskəˈveɪʃn]　发掘，挖掘

respiratory[rəˈspɪrətri]　呼吸的

respirator [ˈrespəreɪtə(r)]　防毒面具，口罩

inspection　检查，视察；检验，审视

rectify [ˈrektɪfaɪ]　改正，矫正

Safety Officer　安全主任

be responsible for　对…负责

working section　施工队

safety organization　安全工作人事结构

the Ministry of Housing and Urban-Rural Development　住房和城乡建设部

jobsite safety policy　现场安全策略

Chinese Labour Law　中国劳动法

safety regulations　安全制度

construction accidents　施工事故

reporting　报告

electrical safety　电气安全

personal safety equipment　个人安全设备

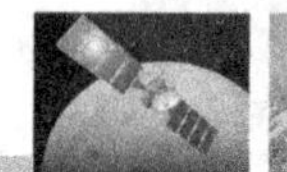

crane safety　起重机安全
respiratory protection　呼吸防护
fire protection and prevention　防火
toxic substances　有毒物质
first aid procedure　急救措施
emergency aid procedure　紧急措施

Chapter 4 Bridge Engineering

第4章 桥梁工程

教学导航

教	知识重点	4.1 General Introduction of Bridge 桥梁概论; 4.2 Bridge Superstructure 桥梁上部结构; 4.3 Bearing 支座; 4.4 Bridge Substructure 桥梁下部结构; 4.5 Bridge Rehabilitation and Consolidation 桥梁维修与加固
	知识难点	桥梁工程相关专业英语词汇的记忆与背诵
	推荐教学方式	明确教学目标，依据教材、多媒体课件、视频等教学手段，让学生掌握桥梁工程相关专业英语知识并且利用科学的单词记忆法引领学生背诵、积累桥梁工程领域的专业英语词汇
	建议学时	12 学时
学	推荐学习方法	以任务驱动、小组讨论、课外拓展的学习方式为主。结合本章内容，课堂学习与自主学习相结合
	必须掌握的理论知识	桥梁工程相关专业英语知识与词汇
	必须掌握的技能	通过对本章的学习，学生能够识读桥梁工程相关专业英语词汇，并能够阅读与桥梁工程相关的英文文献、学术论文等

4.1 General Introduction of Bridge 桥梁概论

扫一扫看本节参考译文

扫一扫看本节教学课件

4.1.1 Text

Definition

A bridge is a structure constructed to provide a passage over an obstacle such as a road, railway, river, and valley. Designs of bridges vary according to the function of the bridge and the nature of the terrain where the bridge is situated. The bridge may be built for road (Fig.4-2, 4-3), railway (Fig.4-1, 4-3), canal, pipeline, cycle track or pedestrians (Fig.4-4).

The field of bridge engineering, which is a branch of civil engineering, includes planning, design, construction, maintenance and rehabilitation.

Fig.4-1 Railway Bridge

Fig.4-2 Highway Bridge

Fig.4-3 Combined Highway and Railway Bridge

Fig.4-4 Footbridge

Selecting the site for a bridge

The factors to be taken into consideration while selecting the bridge site are as follows:

1	The bridge should cross the river at right angles to the direction of river flow so as to minimize the length of the bridge.
2	The banks on both sides of the river should have firm soil and be straight and well-defined to increase the stability of the bridge and reduce the possibility of the erosion of the banks.
3	The bridge site should be at a place where the river is narrow and the flow is streamlined without serious whirls and cross currents to reduce the length of the bridge and to guard against scour.

续表

4	The selected bridge site should be far away from where the river is likely to change the course.
5	Hard nonerodable strata or rock should be close to the river bed level.
6	There should not be any sharp curves in road approaches.

Components of a bridge

Generally, a *bridge*[1] can be divided into two major parts: superstructure and substructure (Fig.4-5).

Superstructure is a structure above the level of bearing, which carries the traffic, including load bearing structure (i.e. beam, girder, truss, cable and arch) and deck system.

Substructure is a supporting system for superstructure. It consists of piers, abutments and foundations for the piers and abutments.

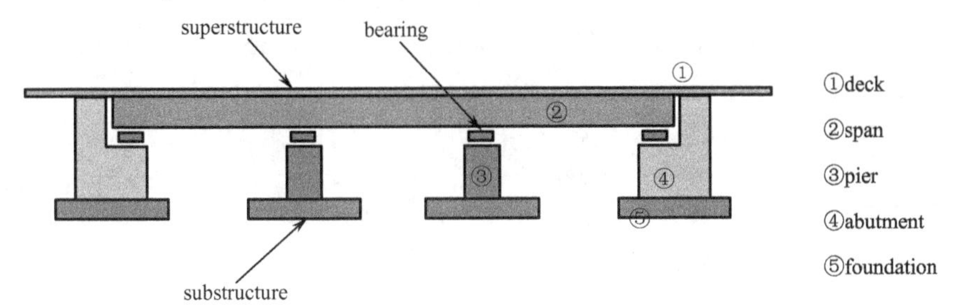

Fig.4-5　Components of a bridge

①**Deck system**　Deck system usually consists of deck, drainage system, expansion joints, footpath, guard stones, handrails and light posts. Sometimes the deck is covered with asphalt concrete or other pavements. Footpaths are provided for pedestrians to walk along without interfering with the heavy vehicular traffic. In order to prevent a vehicle from striking the parapet wall or the hand rails, guard stones painted white are provided along the edge of the footpaths at the ends of the road surface. Handrails are provided on both sides of a bridge to prevent any vehicle from falling off the bridge. (Fig.4-5, Fig.4-6)

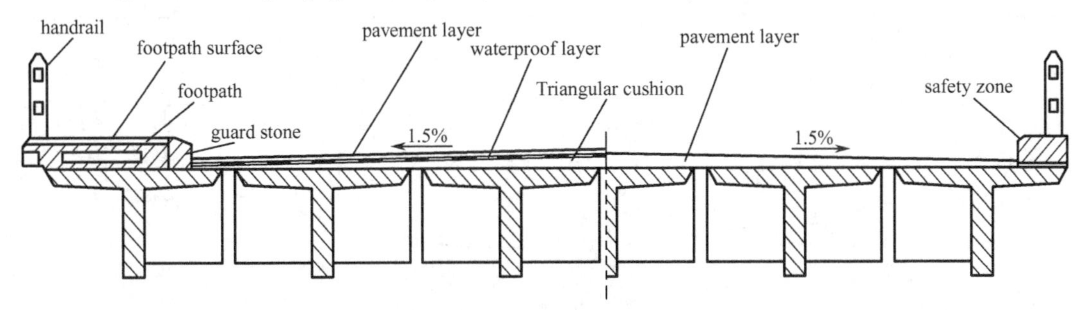

Fig.4-6　Deck systems

②**Span**　Span is the distance between two intermediate supports for a bridge. Span structure is the main load-bearing structure of a bridge. Bridges may be classified by how the forces of tension (suspension bridge), compression (arch bridge), bending (beam bridge), torsion and shear

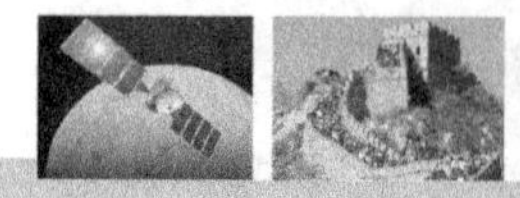

(girder bridge) are distributed through their span structure.

③**Pier** Piers are built between the two end supports of the bridge (abutments) and in the bed of the river to reduce the span and share the total dead load or live load coming from the bridge.

④**Abutment** The end supports of a bridge are called abutments, which may be made of brick masonry, stone masonry, reinforced concrete or precast concrete blocks. It serves both as a pier and as a retaining wall. The height of abutment is equal to height of the piers. The functions of an abutment are as follows:

★ To transmit the load from the bridge superstructure to the foundations;

★ To retain earth work of embankment of the approaches which keep the communication route.

⑤**Foundation** The lowest parts of piers and abutments, which are in direct contact with the subsoil supporting the total loads from the structure are called foundations.

The factors which affect the selection of foundation include the type of soil, the nature of soil, the type of the bridge, the velocity of water and the superimposed load on the bridge.

Technical terms (mainly in Fig.4-7)

No.	technical terms	notes
1	span	It is the center to center distance between two supports.
2	culvert	It is a small bridge having maximum span of 6m.
3	high flood level	It is the level of the highest flood ever recorded in a river or stream.
4	ordinary flood level	It is the flood level which generally occurs every year.
5	low water level	It is the minimum water level in the dry weather.
6	free board	The difference between the high flood level and the level of the crown of the road at its lowest point is called free board.
7	head room	It the highest point of a vehicle or vessel and the lowest point of any protruding member of a bridge.
8	length of the bridge	It will be taken as the overall length measured along the center line of the bridge from the beginning to the end.

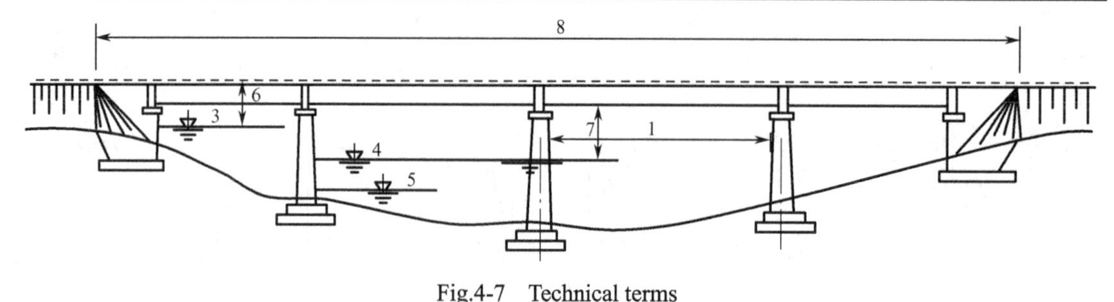

Fig.4-7 Technical terms

Types of bridge

1. According to ***the structural system*** of the bridges, the bridges can be divided into five main types: beam bridges (Fig.4-8), arch bridges (Fig.4-9), suspension bridges (Fig.4-10), rigid frame

bridges (Fig.4-11) and combined system bridges (Fig.4-12).

Fig.4-8　beam bridge

Fig.4-9　arch bridge

Fig.4-10　suspension bridge

Fig.4-11　rigid frame bridge

Fig.4-12　cable-stayed bridge (combined system bridge)

2. According to the ***function or utility*** of the bridges, the bridges are classified into railway bridges (Fig.4-1), highway bridges (Fig.4-2), combined highways and railway bridges (Fig.4-3), footbridges (Fig.4-4) and so on.

3. According to ***the span and length*** of bridges, the bridges can be divided into grand bridges, bridges, middle bridges, small bridges, and culverts. For details, please refer to the specification JTG D60—2015[2].

4. According to ***the position of the floor*** of the bridges, the bridges are classified into deck bridges (Fig.4-13), through bridges (Fig.4-14) and semi-through bridges (Fig.4-15).

Fig.4-13　deck bridge

Fig.4-14　through bridge

Fig.4-15　semi-through bridge

5. According to ***materials*** of bridges, the bridges are classified into timber bridges (Fig.4-16), masonry bridges (Fig.4-17), steel bridges (Fig.4-18), RC (Reinforced Concrete) bridges (Fig.4-19) and prestressed concrete bridges.

Fig.4-16　timber bridge

Fig.4-17　masonry bridge

Fig.4-18　steel bridge

Fig.4-19　RC bridge

6. According to ***life expectancy*** of bridges, the bridges are classified into temporary and permanent bridges.

4.1.2　Notes

1．桥梁是指为使道路跨越天然或人工障碍物而修建的建筑物。总体上说，桥梁由四个基本部分组成，即上部结构（superstructure）、下部结构（substructure）、支座（bearing）和附属设施（accessory），桥梁结构组成如下图所示：

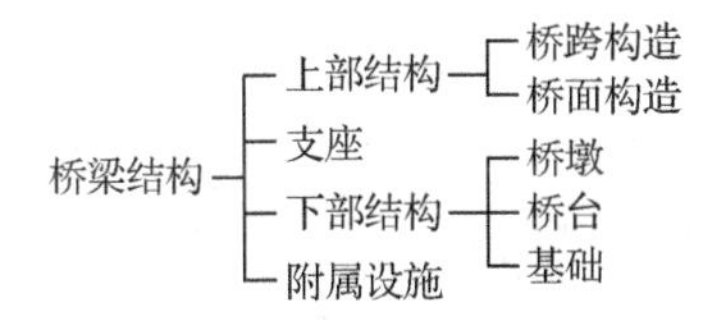

（1）桥跨构造——指桥梁中直接承受桥上交通载荷的、架空的主体结构部分。对梁桥而言，其主体结构是梁；对拱桥而言，其主体结构是拱；对悬索桥而言，其主体结构是缆。

（2）桥面构造——指为保证桥跨结构能正常使用而需要建造的桥上各种附属结构或设施。行车道路面的铺装、人行道、安全带（护栏）、排水防水系统、伸缩装置、路缘石、栏杆、照明等。

（3）支座——在桥跨结构与墩台之间提供载荷的传递途径。

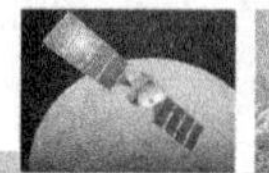

（4）下部结构——指桥梁位于支座以下的部分，也叫支承结构。

桥台：在桥跨结构的两端起支承作用，还起到与路堤衔接、防止路堤滑塌的作用。为此，通常需要在桥台周围设置锥体护坡。

桥墩：分设在两桥台之间。

基础：承受全部载荷并将其传递给地基，因此需要埋入土层之中或建筑在基岩之上。

2．根据《公路桥涵设计通用规范》（JTG D60—2015），特大桥、大桥、中桥、小桥及涵洞按多孔跨径总长或单孔跨径分类的规定如下表所示：

桥涵分类	多孔跨径总长 L（m）	单孔跨径 L_K（m）
特大桥	$L>1\,000$	$L_K>150$
大桥	$100\leqslant L\leqslant 1\,000$	$40\leqslant L_K\leqslant 150$
中桥	$30<L<100$	$20\leqslant L_K<40$
小桥	$8\leqslant L\leqslant 30$	$5\leqslant L_K<20$
涵洞	—	$L_K<5$

4.1.3 New Words and Expressions

obstacle [ˈɒbstəkl]	障碍（物）
valley [ˈvæli]	山谷，溪谷
function [ˈfʌŋkʃn]	功能，作用
terrain [təˈreɪn]	地形，地势
situate [ˈsɪtʃueɪt]	使位于，使处于……地位（位置）
pipeline	管道；输油管道
cycle track	自行车车道
bridge engineering	桥梁工程
branch	分科
civil engineering	土木工程
planning	规划
design	设计
construction	施工
rehabilitation	修复
at right angles to	与……成直角
length of the bridge	桥长
well-defined	定义明确的；界限清晰的
stability [stəˈbɪləti]	稳定（性）
erosion [ɪˈrəʊʒn]	腐蚀，侵蚀
bridge site	桥址
streamlined ['stri:mlaɪnd]	（指汽车、飞机等）流线型的
whirl [wɜ:l]	旋转，回旋

scour [ˈskaʊə(r)]	冲刷
nonerodable	不可蚀的
stratum	地底、岩层
river bed	河床
superstructure	上部结构
substructure	下部结构
bearing	支座
beam	梁
girder [ˈgɜ:də]	大梁、主梁
truss [trʌs]	桁架
cable [ˈkeɪbl]	缆、索
arch [ɑ:tʃ]	拱
deck	桥面
pier [pɪə]	桥墩
abutment [ə'bʌtmənt]	桥台
foundation	基础
span	跨径
deck system	桥面系统
drainage system	排水系统
footpath	人行道
guard stone	缘石
handrail	扶手，栏杆
light post	灯杆
asphalt concrete	沥青混凝土
vehicular [viˈhɪkjələ(r)]	车的，用车辆运载的
parapet wall	护墙，矮墙
road surface	路面
tension [ˈtenʃn]	拉力
compression [kəm'preʃn]	压力
bending ['bendɪŋ]	弯矩
torsion [ˈtɔ:ʃn]	扭转力
shear [ʃɪə(r)]	剪力
suspension bridge	悬索桥
arch bridge	拱桥
beam bridge	梁桥
girder bridge	箱梁桥
dead load	静载荷

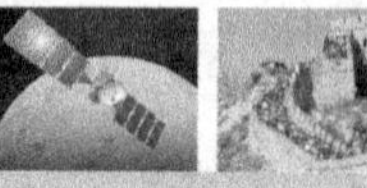

live load	活载荷
brick [brɪk]	砖
masonry [ˈmeɪsənri]	石工工程，砖瓦工工程
reinforced concrete	钢筋混凝土
precast concrete block	预制混凝土试块
earthwork [ˈɜ: θ wɜ:k]	土方（工程）
embankment [ɪmˈbæŋkmənt]	路堤
approach	引道，引路
subsoil [ˈsʌbsɔɪl]	地基土
velocity[vəˈlɒsəti]	速度
superimposed load	附加载荷
high flood level	高水位
designed flood level	设计水位
ordinary flood level	一般水位
low water level	低水位
free board	安全高度
head room	净空高度
vessel [ˈvesl]	船只
protruding [prə'tru:diŋ]	突出的、伸出的
member	构件
center line	中心线
structural system	结构体系
rigid frame bridge	刚构桥
combined system bridge	组合体系桥
utility [ju:ˈtɪləti]	功用，效用
highway bridge	公路桥
railway bridge	铁路桥
combined highway and railway bridge	公铁两用桥
footbridge	人行桥
grand bridge	特大桥
deck bridge	上承式桥
through bridge	下承式桥
semi-through bridge	中承式桥
timber bridge	木桥
masonry bridge	圬工桥，砌体桥
steel bridge	钢桥
RC bridge	钢筋混凝土桥

prestressed concrete bridge	预应力混凝土桥
life expectancy	使用期限
temporary bridge	临时桥
permanent bridge	永久桥
accessory [ək'sesəri]	附属设施

4.2　Bridge Superstructure 桥梁上部结构

扫一扫看本节参考译文

4.2.1　Text

Bridge structural system consists of superstructure and substructure. The upper portion of the bridge structure that directly receives the live load is referred to as the superstructure. Bridge superstructure contains deck system and span. All feasible superstructure types and span arrangements should be considered in the preliminary phases of the project and the relating various researches on the bridge superstructure types and span arrangements also should be done in the bridge design process. Especially, steel and concrete superstructures have many types, and each type has itself structural characteristic and span arrangement. Superstructure types are classified into beam bridge, arch bridge, suspension bridge and combined system bridge as follows according to the structural system.

扫一扫看本节教学课件

Beam bridge

A *beam bridge*[1] is a horizontal, rigid structure that is situated on two end supports, and carries traffic loads by acting structurally as a beam. It evolves from the log bridge (Fig.4-20), and now beam bridges are more built by using reinforced concrete, or prestressed concrete (Fig.4-21).

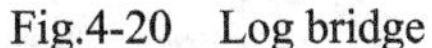
Fig.4-20　Log bridge

Fig.4-21　Prestressed concrete bridge

When a span of beam bridge (Fig.4-22) is supported by only two piers or abutments, it is called a ***simply supported beam bridge***. If two or more beams are joined rigidly together over supports, the bridge becomes ***continuous beam bridge***.

Cantilever beam bridges normally use pairs of cantilevers back to back with a short beam bridge between the cantilevers. Modern motorways often adopt single-cantilever bridge type, which uses a cantilever coming out from each side and a beam bridge between them.

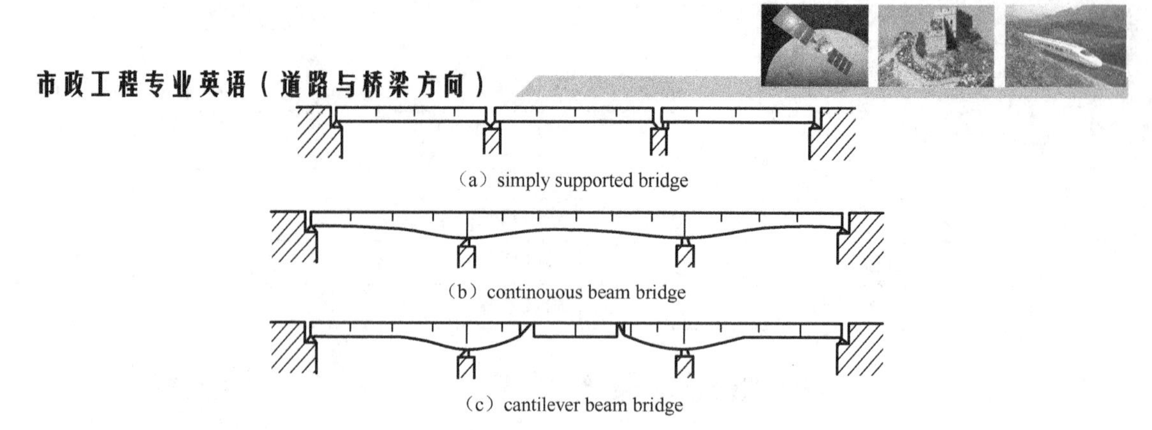

（a）simply supported bridge

（b）continouous beam bridge

（c）cantilever beam bridge

Fig.4-22 types of beam bridge

Bridge beams may be different in cross sectional shape. Below are three kinds of beam shapes (Fig.4-23): a box section, ribbed beam, and plate beam. For a reinforced concrete or prestressed concrete beam, some may be solid or hollow. Solid beams are heavier than hollow beams. Hollow beams are given a special cross section for strength and rigidity. They may be as strong as the solid beams but are much lighter.

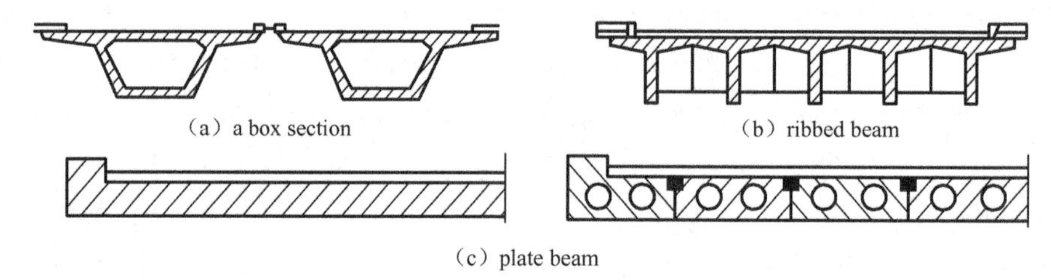

（a）a box section

（b）ribbed beam

（c）plate beam

Fig.4-23 cross sectional shape

Arch bridge

Arch bridge derives its name directly from its shape. Downward force from the top of arch is carried along the curving form to the base. At the same time, the ground pushes back with an equal force. As a result, each of the arch sections is tightly squeezed or compressed by adjacent sections, making the structure very rigid. The ratio of rise to span (f/L) is a very important parameter to reflex the structural characteristics of the arch bridge. (Fig.4-24)

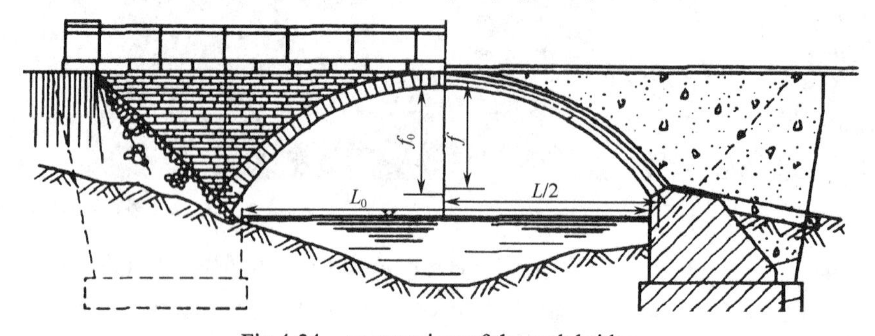

Fig.4-24 an overview of the arch bridge

For deck arch bridges, arches are main structure to bear all the loads on the bridge and then transfer the loads to pries, abutments and foundations. Because of the curvature of the arch, vehicles can not travel smoothly, materials (gravel, crushed stone, pebble, soil and cement) thus are filled as *spandrel structure*[2]. The spandrel structure can be built as solid or hollow and the corresponding arch bridges are known as filled spandrel arch bridges (Fig.4-26) and open spandrel arch bridges (Fig.4-25).

Fig.4-25　Open spandrel arch bridge

Fig.4-26　Filled spandrel arch bridge

Types of arch bridge are shown in the following table:

According to the deck: deck arch bridge, half-through arch bridge and bottom-through arch bridge.
According to the spandrel structure: filled spandrel arch bridge and open spandrel arch bridge.
According to the arch axis: circle arch bridge, catenary arch bridge and parabolic arch bridge.
According to the matarial: stone arch bridge, reinforced concrete arch bridge and steel arch bridge.

Among them, many arch bridge types in construction are ribbed arch bridges (Fig.4-27), curved arch bridges (Fig.4-28), trussed arch bridges (Fig.4-29), and rigid framed arch bridges (Fig.4-30). The majority of these structures is deck bridges with wide clearance and cost less.

Fig.4-27　ribbed arch bridge

Fig.4-28　curved arch bridge

Fig.4-29　trussed arch bridge

Fig.4-30　rigid framed arch bridge

Suspension bridge

Suspension bridge[3] is a kind of bridge that the main load-bearing elements are suspension cables. Well suspended from two high points over a river or canyon, simple suspension bridges (Fig.4-31) form a downward arc within a narrow range, which are not suitable for modern roads and railroads. With the development of the materials and design, the suspended-deck suspension bridge appears. The modern suspension bridge (Fig.4-32), whose cables are suspended between towers and vertical suspenders carrying the weight of the bridge deck and the vehicular loads, provides a steady communication route for vehicles and light rails.

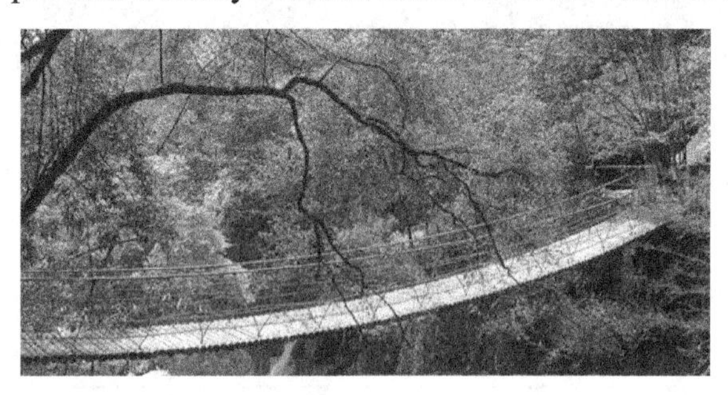

Fig.4-31 simple suspension bridge

Fig.4-32 Golden Gate Bridge

Suspension cable must be anchored at both ends of the bridge because all loads acting on the bridge will be converted into the tension on the main cable. The main cable passes through the towers that is attached to caissons or cofferdams and connected to the deck through suspenders or cables, and then anchored in the foundation (Fig.4-33).The main forces in a suspension bridge are tension on the main cables and compression on the pillars.

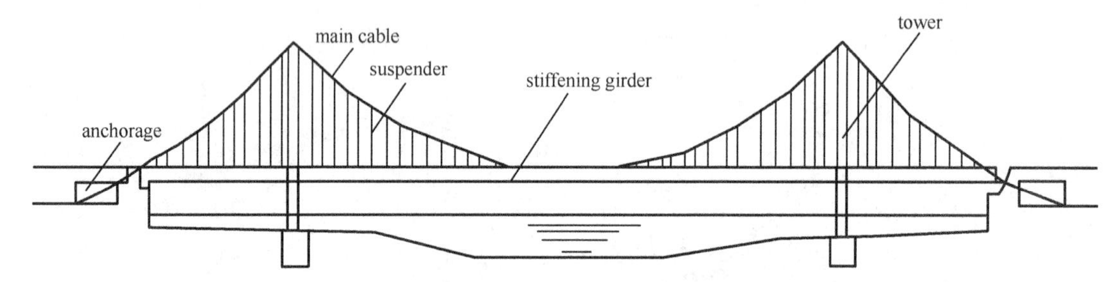

Fig.4-33 main structural members of ground-anchored suspension bridge

Compared with other bridge types, advantages and disadvantages of suspension bridge are listed in the following tables.

advantages	disadvantages
longer main spans	To prevent the bridge deck vibrating under high winds, considerable stiffness should be required or aerodynamics should be taken into account.
less material, lower construction cost	The relatively lower deck stiffness makes it more difficult to carry heavy rail traffic, because high concentrated live loads occur on the deck.
be better able to withstand seismic movements	Some clearance below may be required during construction in order to lift the initial cables or to lift deck units.

Cable-stayed bridge

A *cable-stayed bridge*[4] is a bridge which consists of one or more columns (towers or pylons) and cables supporting the bridge deck. There are three major kinds of cable-stayed bridges, differentiated by how the cables are connected to the tower(s) (Fig.4-34). The cable-stayed design is the optimum bridge for a large span length.

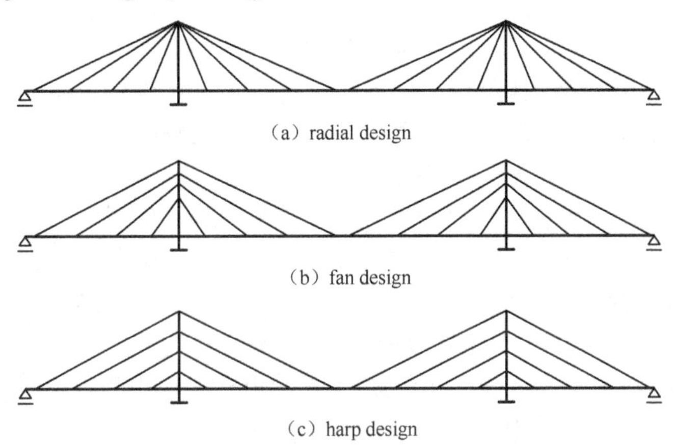

(a) radial design

(b) fan design

(c) harp design

Fig.4-34　cable-stayed bridge

In the cable-stayed bridge, the towers are the primary load-bearing structures. A cantilever approach is often used for supporting the bridge deck near the towers, but further bridge decks are often supported by cables running directly to the towers. Compared to the suspension bridge, the disadvantage of the cable-stayed bridge is that the cables pull to the sides as opposed to directly up, so the bridge deck should be stronger to resist the resulting horizontal compression loads; but this kind of bridge has the advantage of not requiring firm anchorages to resist a horizontal pull of the cables, as in the suspension bridge. All the static horizontal forces are balanced so that the supporting tower does not tend to tilt or slide, needing only to resist such forces from the live loads (Fig.4-35).

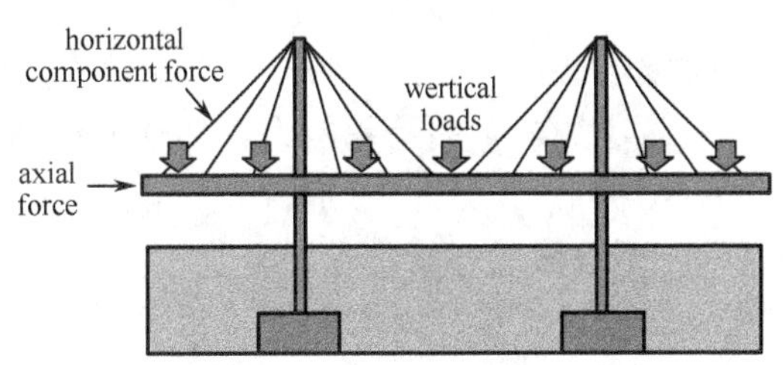

Fig.4-35　the force analysis of the cable-stayed bridge

Main advantages of the cable-stayed bridge are as follows:

1. The cable-stayed bridge has much greater stiffness than the suspension bridge, so that deformations of the deck under live loads are reduced.

2. It can be constructed by cantilevering out from the piles the cables act both as temporary and permanent supports to the bridge deck.

3. It is a symmetrical bridge (i.e. spans on either side of the tower are the same); the horizontal forces balance and large ground anchorages are not required.
4. Any number of towers may be used in the cable-stayed bridge. This bridge type can be as easily built with a single tower as with a pair of towers. However, a suspension bridge is usually built only with a pair of towers.

4.2.2 Notes

1．梁桥结构从力学特点上，可以分为简支梁桥、连续梁桥和刚构梁桥三类。

（1）简支梁桥：桥跨结构在墩台处断开，不连续，各跨结构之间的受力和变形相互无影响。简支梁：受力明确，构造简单，适用范围广，不受基础条件限制，易于建造、标准化和更换，适用的跨度范围一般在 20 米以下。

（2）连续梁桥：桥跨结构在墩台处连续，各跨结构之间的受力和变形相互有影响。

（3）刚构梁桥：除了桥跨结构与桥墩在墩台处连续外，梁和墩（台）构成刚性连接。连续梁桥和刚构梁桥与简支梁桥相比较，可以减小跨内的弯曲变形，降低主梁的高度，从而减少材料用量和结构自重，适用的跨度范围为 25 米以上。

2．实腹拱：拱圈上部的建筑是实心的，一般采用石、砂等建筑材料为媒介，将上部载荷均匀传递给拱圈。实腹拱的构造简单、自重大，适用于中、小跨度。

空腹拱：拱圈与桥面板之间采用柱列式结构相连，将上部载荷均匀传递给拱圈。空腹拱的结构合理、自重较小、利于泄洪，是大、中跨拱桥常用的形式。

3．悬索桥由主缆（main cable）、加劲梁（stiffening girder）、塔柱（tower）、锚碇（anchorage）构成。与拱的受力特性相似，一般来说，索不承受弯矩和剪力，只有轴力，全截面受拉。较之梁以受弯为主，索能更有效地发挥截面全体材料的承载能力。同其他的桥相比，跨度越大时，悬索桥的优势越明显。主缆作为主要承重构件，承受桥的恒载荷和活载荷，具有非常合理的受力形式。加劲梁即主梁只是传力部件，在材料用量和截面设计方面，其截面积并不需要随着跨度增大而增加。在构件设计方面，悬索桥的主缆、锚碇和塔柱这三项主要承重构件在扩充其截面积或承载能力方面，所遇到的困难较小；在施工方面，风险较小。

4．斜拉桥由从主塔柱往两边伸出的拉索将主梁拉起。若是由多条拉索分散拉起的，主梁就像支撑在多个弹性支座的连续梁一样工作，减小了主梁的跨度，也可以大大降低梁高，减轻自重，减小内力，从而增加桥梁的跨越能力。塔柱的造型和拉索的布置形成不同的桥型风格。斜拉桥的塔柱有 A 字形、门形、H 形、多层框架形、倒 Y 形、独柱形、立柱形等，形式变化很大，且其构造由拉索的布置决定。

斜拉桥在横桥向的布置有三种形式：单索面、双平行索面、双斜索面；在顺桥向的布置有三种形式：辐射形（radial）、扇形（flabellate）、竖琴形（harp）。

根据拉索的锚固方式不同，斜拉桥可分为自锚式、地锚式和部分地锚式三种结构体系。

4.2.3 New Words and Expressions

feasible[ˈfi:zəbl]	可行的；行得通的
span arrangement	跨径布置
in the preliminary phase of project	工程初期

rigid[ˈrɪdʒɪd]	刚性的
log [lɒg]	原木
reinforced concrete	钢筋混凝土
prestressed concrete	预应力混凝土
simply supported beam bridge	简支梁桥
continuous beam bridge	连续梁桥
cantilever beam bridge	悬臂梁桥
cantilever [ˈkæntɪli:və]	（桥梁或其他构架的）悬臂，悬桁，伸臂
motorway	高速公路；汽车道；快车道
cross section	横截面
box beam	箱梁
ribbed beam	肋梁
plate beam	板梁
solid beam	实心梁
hollow beam	空心梁
strength	强度
rigidity	刚度
downward	向下的
squeeze [skwi:z]	挤压，施加压力
compress [kəmˈpres]	压紧，压缩
adjacent [əˈdʒeɪsnt]	邻近的，毗邻的
ratio of rise to span	矢跨比
parameter [pəˈræmɪtə]	【数】参数
deck arch bridge	上承式拱桥
load	载荷
arch	拱圈
spandrel ['spændrəl]	拱肩
spandrel structure	拱上结构，拱上建筑
filled spandrel arch bridge	实腹拱桥
open spandrel arch bridge	空腹拱桥
half-through arch bridge	中承式拱桥
bottom-through arch bridge	下承式拱桥
arch axis	拱轴线
circle arch bridge	圆弧拱桥
catenary arch bridge	悬链拱桥
parabolic arch bridge	抛物线拱桥
stone arch bridge	石拱桥

reinforced concrete arch bridge	钢筋混凝土拱桥
steel arch bridge	钢拱桥
ribbed arch bridge	肋拱桥
curved arch bridge	双曲拱桥
trussed arch bridge	桁架拱桥
rigid framed bridge	刚构拱桥
clearance[ˈklɪərəns]	净空，空隙
suspend [səˈspend]	悬挂
arc [ɑ:k]	弧，弧度
tower	塔，塔柱，索塔
suspender [səˈspɛndɚ]	吊杆
vehicular load	行车载荷
light rail	轻轨
anchor [ˈæŋkɚ]	锚固，固定于
convert [kənˈvɜ:t]	转变，转化
main cable	主缆
caisson ['keɪsən]	沉箱
cofferdam ['kɒfədæm]	沉箱，围堰
pillar [ˈpɪlə]	柱，梁，墩
main span	主跨
withstand [wɪðˈstænd]	经受，承受
seismic [ˈsaɪzmɪk]	地震的，由地震引起的
vibrate [vaɪˈbreɪt]	（使）振动
stiffness [stɪfnəs]	刚度
aerodynamics [ˌeərəʊdaɪˈnæmɪks]	空气动力学
cable-stayed bridge	斜拉桥
column [ˈkɒləm]	柱，梁
pylon [ˈpaɪlən]	（架高压输电线的）电缆塔
differentiate [ˌdɪfəˈrenʃieɪt]	区分，区别，辨别
large span length	大跨度
radial design	辐射形
fan design	扇形
harp design	竖琴形
axial force	轴向力
horizontal component force	水平分力
vertical load	竖向载荷
anchorage [ˈæŋkərɪdʒ]	锚固装置

tilt[tɪlt]	倾斜
slide [slaɪd]	滑落
symmetrical [sɪˈmetrɪkl]	对称的
ground anchorage	地锚结构

4.3 Bearing 支座

扫一扫看本节参考译文

扫一扫看本节教学课件

4.3.1 Text

Bearing[1] is a main part in bridge structures, which is positioned between the bridge superstructure and substructure. The key functions are to transmit loads from the superstructure to the substructure and to accommodate relative movements. The movements in the bridge bearings include **translations** and **rotations** (Fig.4-36). Creep, shrinkage, and temperature effects are the most common causes of the translational movements, which can occur in both transverse and longitudinal directions. Traffic loading, construction tolerances, and uneven settlement of the foundation are the common reasons of the rotations.

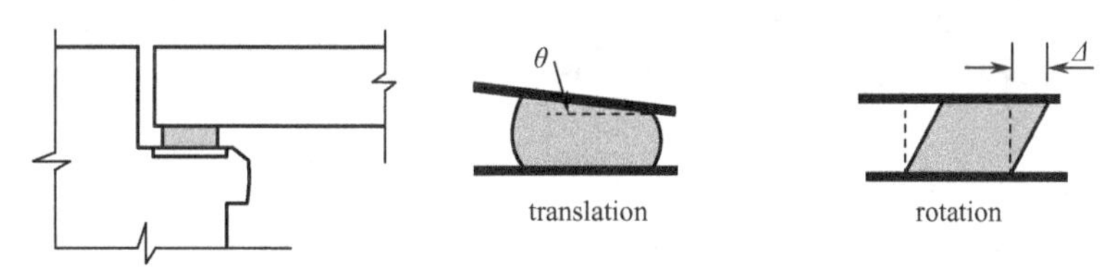

Fig.4-36　movements in the bridge bearings

To determine the suitable bearing types based on the bridge, the vertical and horizontal loads, the rotational and translational movements caused by dead and live loads, wind loads, earthquake loads, creep and shrinkage, prestress, thermal and construction tolerances should be calculated.

Usually, bearing is connected to the superstructure through the use of a steel sole plate and sits on the substructure through a steel masonry plate. The sole plate distributes the concentrated bearing reaction to the superstructure. The masonry plate distributes the reactions to the substructure. For steel girders, the connection between the sole plate and the superstructure is done by bolting or welding. For concrete girders, the sole plate is pre-embedded into the concrete with anchor studs. The masonry plate is typically connected to the substructure with anchor bolts (Fig.4-37).

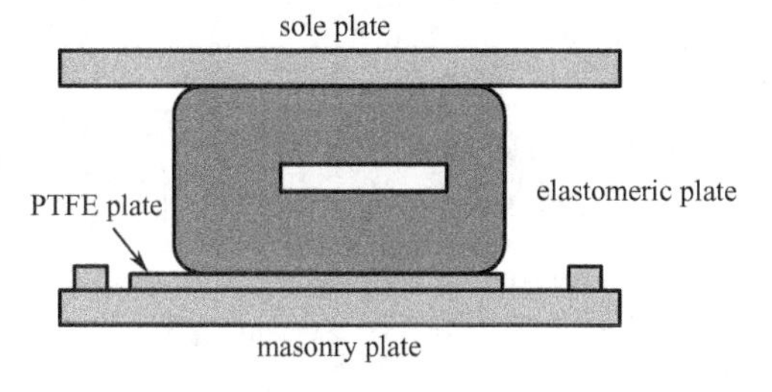

Fig.4-37　bearing

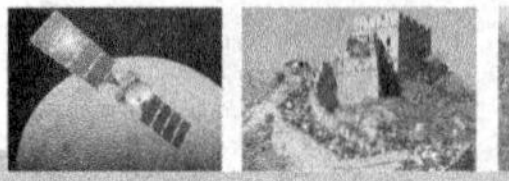

Bearings are classified into fixed bearings and expansion bearings. Fixed bearings allow rotations but restrict translational movements. Expansion bearings allow both rotational and translational movements. There are many types of bearing available. The following table lists the main types of bearing currently in use.

types of bearing	characteristics	illustration
rocker bearing	A rocker bearing is a kind of expansion bearing that appears in a great variety. It typically consists of a pin at the top that facilitates rotations, and a curved surface at the bottom that accommodates the translational movements. Rocker bearings are used in steel bridge for a bridge with a medium span.	Top plate Pin Anchor bolt Slotted hole Masonry plate
pin bearing	A pin bearing is a type of fixed bearing that accommodates rotations through the use of a steel pin. The typical configuration of the bearing is as the same as the rocker described above except that the bottom curved rocker plate is now flat and directly anchored to the concrete pier.	
roller bearing	A roller bearing is composed of one or more rollers between two parallel steel bars. Single roller bearings can facilitate both rotations and translations in the longitudinal direction, while a group of rollers only accommodate longitudinal translations. Roller bearings have been used in both steel and concrete bridges.	Tie bars Rollers Masonry plate
elastomeric bearing	An elastomeric bearing is made of elastomer. It accommodates both translational and rotational movements through the deformation of the elastomer. Elastomer is flexible in shear but very stiff against volumetric change. Under compressive load, the elastomer expands laterally. To sustain large load without excessive deflection, reinforcement is used to restrain lateral bulging of the elastomer. This leads to the development of several types of elastomeric bearing pad.	Sole plate Stainless steel surface PTEF sliding surface Elasomeric pad Masonry plate Steel reinforement Elastomer

续表

types of bearing	characteristics	illustration
pot bearing	A pot bearing comprises a plain elastomeric disk that is confined in a shallow steel ring, or pot. Vertical loads are transmitted through a steel piston that fits closely to the steel ring. Flat sealing rings are used to contain the elastomer inside the pot. The elastomer behaves like a viscous fluid within the pot as the bearing rotates. Because the elastomeric pad is confined, much larger load can be carried this way than through conventional elastomeric pads.	Sole plate Keeper plate PTEF sliding surface Steel piston Elasomeric pad Sealing ring Pot
spherical bearing	A curved bearing consists of two matching curved plates with one sliding against the other to accommodate rotations.	Sole plate Upper concave plate PTEF sliding surface Lower convex plate Masonry plate
disk bearing	A disk bearing uses a PTFE disk to support the vertical loads and a metal key in the center of the bearing to resist horizontal loads. The rotational movements are accommodated through the deformation of the elastomer. To accommodate translational movements, a PTFE slider is required.	Sole plate Upper steel plate PTEF sliding surface Masonry plate Elastmeric disk Stcel key

4.3.2 Notes

1．支座系统是设置在桥梁上、下结构之间的传力和连接装置。其作用是传递上部结构的支承反力，把上部结构的载荷传递到墩台上，保障结构在活载荷变化、温度变化、混凝土收缩和徐变等因素作用下的自由变形，以使上、下部结构的实际受力情况符合结构的静力要求，示意图如下所示。

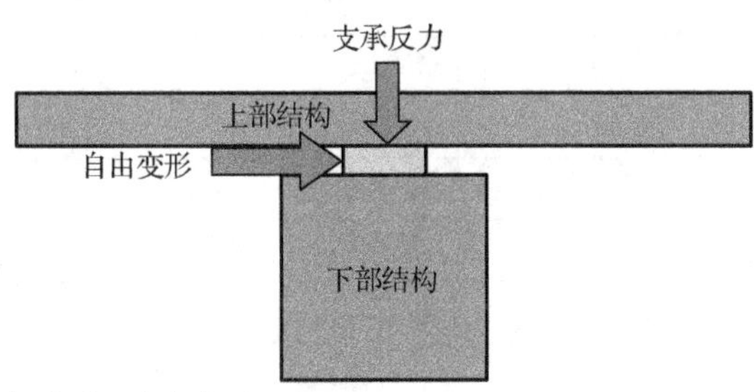

支座一般分为固定支座和可动支座。

4.3.3 New Words and Expressions

bearing [ˈbeərɪŋ]	支座
transmit [trænsˈmɪt]	传递，传送
accommodate [əˈkɒmədeɪt]	适应
translation [trænsˈleɪʃn]	平移

rotation [rəʊˈteɪʃn]	转动
creep [kri:p]	徐变
shrinkage[ˈʃrɪŋkɪdʒ]	收缩
translational movement	平移
transverse [ˈtrænzvɜ:s]	横向的
longitudinal direction	竖向
traffic load	行车载荷
construction tolerance	施工误差
uneven settlement of the foundation	地基不均匀沉降
earthquake load	地震载荷
prestress ['pri:'stres]	给……预加应力
thermal [ˈθɜ:ml]	热的
sole plate	底板，上支座板
masonry plate	座板，下支座板
reaction [riˈækʃn]	反力，反作用力
girder [ˈgɜ:də]	大梁
bolt [bəʊlt]	螺栓
weld [weld]	焊接
embed [ɪm'bed]	把……嵌入
anchor stud	锚固螺栓
anchor bolt	锚栓
PTFE(polytetrafluoroethylene)	聚四氟乙烯
elastomeric plate	橡胶板
fixed bearing	固定支座
expansion bearing	可动支座，活动支座
rocker bearing	摇轴支座
pin bearing	铰支座
roller bearing	辊轴支座
elastomeric bearing	橡胶支座
spherical bearing	球形支座
curved bearing	曲面支座，球形支座
pot bearing	盆式支座
disk bearing	板式支座
pin [pɪn]	销钉，铰
facilitate [fəˈsɪlɪteɪt]	有助于，促进，助长
configuration [kənˌfɪgəˈreɪʃn]	布局，构造
anchor [ˈæŋkə]	n.锚；v.锚固于
roller	辊轮

parallel	平行的
elastomer	橡胶
deformation	变形
shear	剪力
stiff [stɪf]	坚硬，刚度
volumetric change	体积变化
compressive load	压力载荷
expand [ɪk'spænd]	扩展；发展；张开；展开，扩大
lateral ['lætərəl]	侧面的，横向的
sustain [sə'steɪn]	维持；支撑，支持
deflection [dɪ'flekʃn]	偏斜，偏移，歪曲；偏斜度；挠度
bulge[bʌldʒ]	膨胀
bearing pad	承重垫片
elastomeric disk	橡胶板
confine [kən'faɪn]	限制；局限于；禁闭；管制；界限，范围
piston ['pɪstən]	【机】活塞
sealing ring	密封圈
viscous ['vɪskəs]	黏性的；半流体的；黏滞的
concave [kɒn'keɪv]	凹面的，凹面
convex ['kɒnveks]	凸面的

4.4　Bridge Substructure 桥梁下部结构

扫一扫看本节参考译文

扫一扫看本节教学课件

4.4.1　Text

The substructure of a bridge is analogous to the walls, columns of a single storey building and foundations supporting it. In a bridge, substructure is a supporting system for the superstructure. It comprises piers, abutments, foundations and wingwalls (mainly in Fig.4-38).

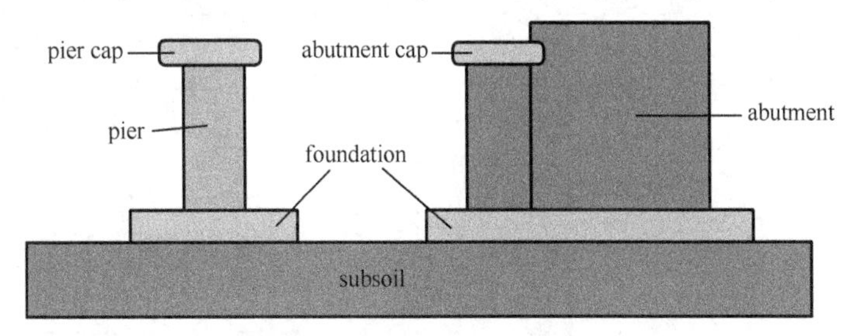

Fig.4-38　Bridge Substructure

Single-span bridge has abutments at each end which provide vertical and lateral support for the bridge and act as retaining walls to resist lateral movement of the earthen fill of the bridge approach. Multi-span bridge requires piers to support ends of spans unsupported by abutments.

Foundation types depend mainly on the depth and safe bearing capacity of the bearing stratum,

also are restricted by differential settlement due to the type of bridgc deck. Bridge foundations are generally split into three categories: spread footings, pile foundations and drilled caisson foundations.

Pier

Piers provide vertical supports for spans at intermediate points of bridge and perform two main functions, that is, transferring superstructure vertical loads to the foundations and resisting horizontal forces acting on the bridge. Although piers are traditionally designed to resist vertical loads, it is becoming more and more common to design piers to resist high lateral loads caused by seismic events, wind loads and impacts of ships. Even outside the earthquake belt, bridge designers are paying more attention to the ductility aspect of the design and the durability of the construction materials. Piers are primarily constructed with reinforced concrete. Steel, to some extent, is also used for piers. Steel tubes filled with concrete columns have gained more attention recently. Pier caps are usually haunched in the region beyond the face of the exterior column or stem. The pier consists of pier cap, pier body and foundation (Fig.4-39).

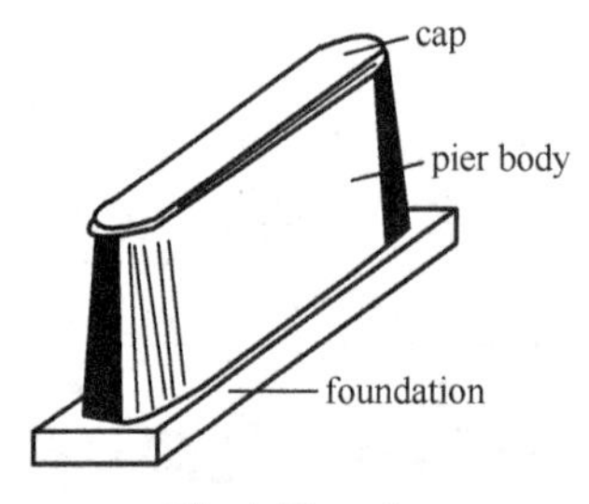

Fig.4-39 pier

Pier[1] is usually used as a general term for any type of substructure located between horizontal spans and foundations. A pier that comprises multiple columns is often called a bent.

There are several ways to classify pier types as shown in the following table.

criteria	types of pier	illustration
the structural connectivity to the superstructure	monolithic pier	Monolithic
	cantilevered pier	cantilever column

续表

criteria	types of pier	illustration
the sectional shape	solid/hollow pier	
	round/octagonal/hexagonal/rectangular pier	
the framing configuration	single/multiple column bent	
	hammerhead/solid wall pier	

Selection of proper pier type depends upon many factors as follows.

First, it depends upon the type of superstructure. Steel girder superstructures are normally supported by cantilevered piers, whereas the cast-in-place concrete superstructures are normally supported by monolithic piers.

Second, it depends upon whether the bridges are over a waterway or not. Pier walls are preferred on river crossings, where hydraulics dictates them multiple column pier type. Multiple pile bents are commonly used on slab bridges.

Last, the height of piers also affects the type selection of piers. The taller piers often adopt hollow cross sections in order to reduce the weight of the substructure which then reduces the load demands on the costly foundations.

Abutment

Abutments are structures positioning at the beginning and end of a bridge, which support the superstructure and approach roadway and retain the earth embankment. The main parts of abutments are the abutment cap, body/stem, and foundation (Fig.4-40). All reinforcing steel throughout the abutment should be epoxy coated.

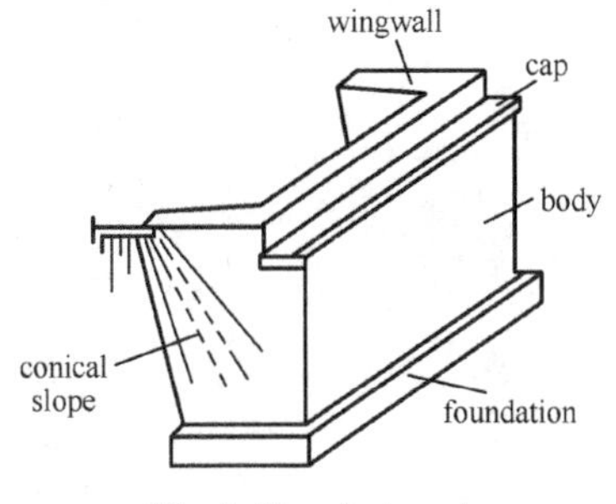

Fig.4-40　abutment

Generally, abutments[2] can be classified into the following four types:

types of abutment	illustration
1. *Gravity abutment or wall abutment* resists horizontal earth pressure with its own dead weight.	Dirt wall Bridge Seat Breast wall River bed
2. *Counterfort abutment*: Abutment with counterfort is used only for very tall walls of more than 10 to 20 metres.	Deck Bearing Stem Water Counterfort wall Foundation
3. *Cellular abutment* has cells which can allow drainage of water. These types of abutment are expensive.	Bearing Water Cell Wall Foundation
4. *Reinforced T-abutment* is slender structure and has a large heel and a small toe. This is a reinforced concrete structure. Reinforced T- abutment is the most common used form of abutment.	Gap for Soft material Bearing Water Abutment Foundation

Foundation

The design of foundation mainly comprises the following stages:

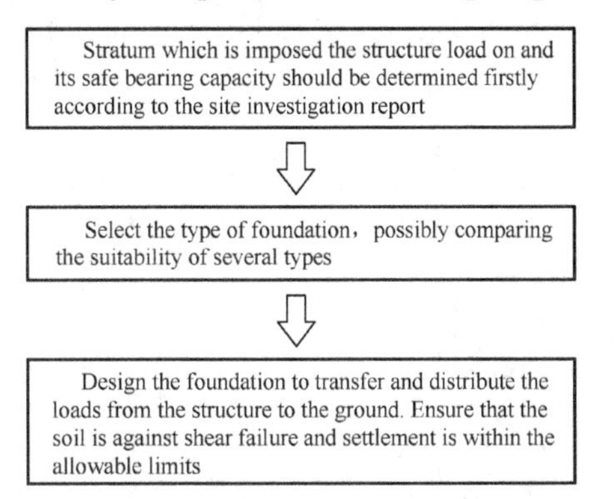

Bridge foundations generally fall into three categories:

1. ***Spread footing*** Spread footings are economical within 20 FT depth. Spread footings should be proportioned to distribute the total vertical and horizontal loads in such a manner that the required structural stability is obtained and that the allowable design bearing pressures are not exceeded (Fig.4-41).

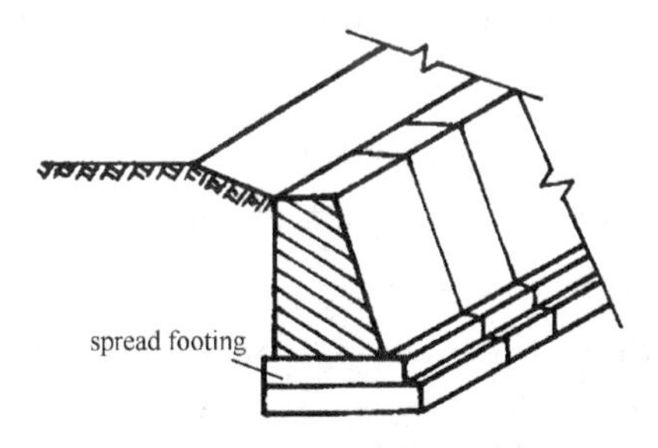

Fig.4-41　spread footing

Spread footings are to be founded on competent rock.Minimum thickness for a spread footing is 3 FT and should be inserted into rock in a minimum of 1 FT. Allowable bearing pressures should be used to size the footing.

2. ***Piled foundation*** Piled foundation must be designed for both axial and lateral loads as appropriate. Loads may include external (non-structural loads) as well as structural loads. For example, piled foundations might be used to enhance stability of the approach embankment if the embankment factor of safety is questionable (Fig.4-42).

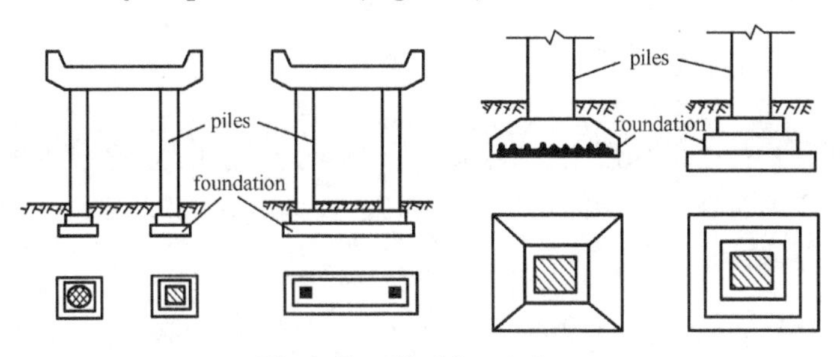

Fig.4-42　Piled foundations

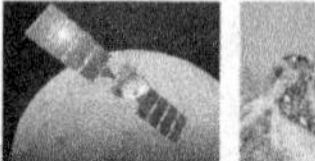

The types of pile generally used for bridge foundations are:

(1) Driven Piles, for example, precast concrete piles or steel piles, etc, may be either end bearing piles or frictional piles, which depends on the strata (Fig.4-43).

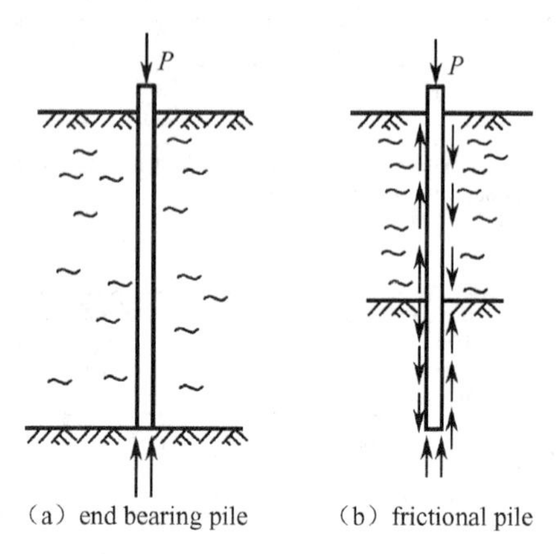

Fig.4-43 driven piles

(2) Driven Cast In-Situ Piles, which are formed by driving a hollow steel tube with a closed end and filling the tube with concrete, or formed by driving a hollow steel tube with a closed end and filling the tube with concrete, simultaneously drawing the tube out.

(3) Bored Piles, which are formed by boring a hole and filling it with concrete, are generally end bearings and often of large diameter. To increase their bearing capacity, the bottom can be under-reamed to produce a greater bearing area.

3. ***Drilled caisson foundation*** Drilled caissons belong to the deep foundations and should be designed in accordance with the specifications and accommodate vertical and lateral loads, which provide superior scour protection versus traditional steel piling, greater resistance against high lateral and uplift loads. Compared with the pile driving process (vibrations, interference of battered piles), the drilled caisson is more suitable for the in-situ conditions. In some cases, cofferdams go out of use. Caisson caps may be eliminated for single or multiple column piers. If in-situ conditions dictate that temporary casings may be uneconomical, the permanent steel casings should be considered to be used.

4.4.2 Notes

1. 桥墩

主要作用：

（1）承受上部结构传来的作用力，并通过基础将此作用力及本身自重传递到地基上。

（2）承受流水压力、风力、地震作用力、冰压力、船只撞击力等。

（3）衔接两岸路线，承受侧岸的附加侧压力。

桥墩的分类如下：

- 桥墩
  - 重力式桥墩
    - 梁桥：墩帽、墩身。
    - 拱桥：墩座、墩身。
    - 特征：依靠自身重力来平衡外力，从而保证桥墩的强度和稳定性。
    - 缺点：数量大、自重大，对地基承载力要求高。
  - 轻型桥墩
    - 梁桥
      - 钢筋混凝土薄壁桥墩：截面积小，自重轻，结构刚度和强度好，多用于高桥，内部是空心的。
      - 柱式桥墩：由分立的两根或多根立柱组成，外形美观，线条简洁明快，体积和自重小。
      - 柔性排架桥墩：可以通过一些连接构造措施将上部结构传来的水平力传递给全桥各个柔性墩台。
    - 拱桥
      - 带三角杆件的单向推力墩。
      - 悬臂式单向推力墩。

2. 桥台

桥台由台帽、台身和基础三部分组成，其分类如下：

- 重力式桥台：主要靠自重来平衡外载荷，以保持自身的稳定性。
- 轻型桥台：利用钢筋混凝土结构的抗弯能力来减少圬工体积而使桥台轻型化。
- 框架式桥台：框架式桥台是一种在横桥向呈框架式结构的桩基础轻型桥台，以框架式结构本身的抗弯能力来减少自重。
- 组合桥台：为使桥台轻型化，桥台本身要承受桥跨结构传来的竖向力和水平力，而台后的土压力则由其他桥跨结构来承受。

1）重力式桥台

组成：由台身（前墙）、台帽、基础与两侧的翼墙组成，在平面上呈 U 字形。

特点：这种桥台的构造简单，但台身较高时工程量较大，一般用于跨度较小的低矮桥台。

2）轻型桥台

分类：薄壁轻型桥台和支撑梁轻型桥台。

优点：结构自重轻，施工方便。

适用范围：小跨径桥梁，桥跨孔数与轻型桥墩配合使用时不宜超过 3 个，单孔跨径不大于 13 m，多孔全长不大于 20 m。

3）框架式桥台

分类：分为柱式、肋墙式、半重力式、双排架式、板凳式。

设置方式：均采用埋置式，台前设置溜坡。为满足桥台与路堤的连接，在台帽上部设置耳墙，必要时在台帽前方两侧设置挡板，以防溜坡土进入支座。

适用范围：适用于地基承载力较小、台身较高、跨径较大的桥梁。

4）组合桥台

分类：分为锚碇板式组合桥台，过梁式、框架式组合桥台，桥台与挡土墙组合桥台。

4.4.3 New Words and Expressions

analogous [əˈnæləgəs]	相似的
single storey building	单层建筑物

wingwall	翼墙
pier cap	墩帽
single-span bridge	单跨桥梁
earthen fill	填土
multi-span bridge	多跨桥梁
bearing stratum	持力层
bearing capacity	承载力
settlement	沉降
spread footing	扩大基础
pile foundation	桩基础
drilled caisson foundation	钻孔沉箱基础
earthquake belt	地震带
ductility [dʌk'tɪlɪtɪ]	展延性，柔软性
durability [ˌdjʊərə'bɪlətɪ]	耐久性；持久性
haunch [hɔ:ntʃ]	n.腰腿（肉）；v.加腋（指在整体结构的转角处同时加大两个相交截面的面积，一般做成三角形，与相交的结构同时浇筑。加腋部分应适当配置构造钢筋。）
stem [stem]	墩身，台身
multiple [ˈmʌltɪpl]	多重的；多个的
bent	排架
criteria[kraɪˈtɪərɪə]	标准，准则（criterion 的名词复数）
illustration	图示，图解
connectivity	连接性
monolithic pier	整体式桥墩
cantilevered pier	悬臂式桥墩
sectional shape	截面形状
solid pier	实心式桥墩
hollow pier	空心式桥墩
octagonal[ɒk'tægənl]	八角形的，八边形的
hexagonal[heks'ægənl]	六角形的，六边形的
rectangular [rek'tæŋgjələ]	矩形的
framing configuration	框架结构
hammerhead pier	锤头式桥墩
cast-in-place concrete	现浇混凝土
waterway [ˈwɔ:təweɪ]	航道
hydraulics [haɪˈdrɔ:lɪks]	水力学
slab bridge	板桥

approach roadway	引道，引桥
retain [rɪˈteɪn]	保持
abutment cap	台帽
epoxy[ɪˈpɒksi]	环氧的；环氧树脂
coat	涂料层；覆盖层
conical slope	锥形护坡
gravity abutment	重力式桥台
dead weight	静负载，固定负载
counterfort　[ˈkaʊntəfɔ:t]	扶壁
slender [ˈslendə]	苗条的，薄弱的
heel [hi:l]	脚后跟
toe [təʊ]	脚尖
suitability [ˌsju:tə'bɪlətɪ]	合适，适合
distribute	分布
shear failure	剪力破坏
within the allowable limits	在允许的范围内
competent rock	强岩层
insert [ɪnˈsɜ:t]	插入，嵌入
axial load	轴向载荷
lateral load	横向载荷
piled foundation	桩基础
enhance [ɪnˈhɑ:ns]	提高，增加；加强
driven pile	灌注桩
precast [ˌpri:ˈkɑ:st]	预浇铸的，预制的
end bearing pile	端承桩
frictional pile	摩擦桩
strata [ˈstrɑ:tə]	地层；岩层（stratum 的名词复数）
driven cast in-situ pile	沉管灌注桩
steel tube	钢管
simultaneously[ˌsɪməlˈteɪnɪəslɪ]	同时地
draw [drɔ:]	拉，拔出
bored pile	钻孔灌注桩
diameter[daɪˈæmɪtə]	直径
ream [ri:m]	扩展，扩张
deep foundation	深基础
in accordance with	与……一致
specification [ˌspesɪfɪˈkeɪʃn]	规范

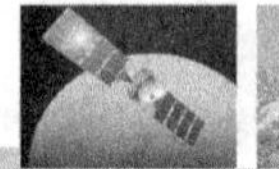

vertical [ˈvɜːtɪkl]	垂直的，竖立的
lateral [ˈlætərəl]	侧面的；横向的
superior [suːˈpɪəriə]	（级别）较高的；（质量）较好的；（数量）较多的；
versus [ˈvɜːsəs]	与……相对；对抗
vibration [vaɪˈbreɪʃn]	振动
battered pile	斜桩
caisson cap	沉井帽
temporary casing	临时护筒
permanent steel casing	永久钢护筒

4.5 Bridge Retrofit and Reinforcement 桥梁维修与加固

4.5.1 Text

扫一扫看本节参考译文

扫一扫看本节教学课件

Introduction

From 2000 to 2013, the numbes of bridges in china increased rapidly (Fig.4-44) By the end of 2016, there were 1 million bridges all over the country. Simultaneously, bridge infrastructure in China is deteriorating at an alarming rate. Due to the staggeringly high cost of repair and replacement, most transportation agencies are unable to cope with this trendency. By the end of 2006, nearly 6 282 bridges in China had been classified as deficient. Structural deficiency does not imply that a bridge is unsafe or likely to collapse. More than 3 000 bridges are classified as structurally or functionally deficient because of poor deck conditions, lack of load ratings or weight restrictions. Much of the deck deterioration can be attributed to heavy application of road de-icing salts during the snowy winters experienced in the nation. The Ministry of Transport estimates that repairing deficient or obsolete bridges will cost more than 20 billion RMB. During the period of "the 10th five-year plan", the Ministry of Transport spent about 15 billion RMB on the more than 7 000 dangerous bridge.

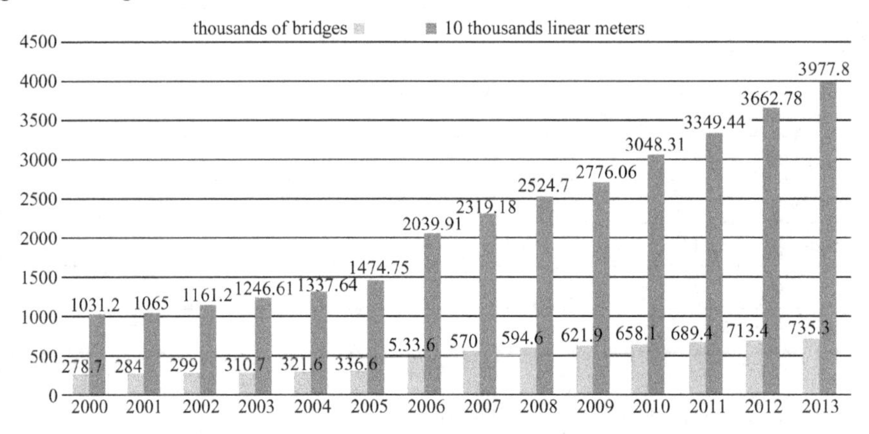

Fig.4-44 National bridge scale increasing curve, 2000-2013

Retrofit and reinforcement

The Ministry of Transport is constantly looking for new materials, methods and technologies to cost-effectively replace old bridge decks and improve load ratings. Fiber reinforced polymer (FRP) composite systems are one of such alternatives. Fiber reinforced polymers are gaining popularity in the bridge community. These materials have high strength-to-weight ratios and excellent durability against corrosion of materials. They have a long record of application in Europe and America. China has recently begun using and evaluating FRPs for repairing bridge deck, strengthening deteriorated components, removing load postings, and prolonging service life.

China has many old truss bridges with deteriorated superstructures, and these bridges are restricted to less than legal loads. Due to the nature of these bridges, replacement is often a cost effective option. Since resources are limited, many of these bridges may not be replaced with new structures for several years. FRP decks seem to offer a cost-effective replacement since they are much lighter than conventional bridge decks (Fig.4-45, Fig.4-46). FRP decks not only replace deteriorated bridge decks, but also reduce the dead load. The allowable live load capacity increases with the dead load reduction and the retrofited bridges can carry legal loads without extensive repairs. Installation is relatively fast, reducing the inconvenience to the traveling public.

Fig.4-45　old bridge retrofited by FRP

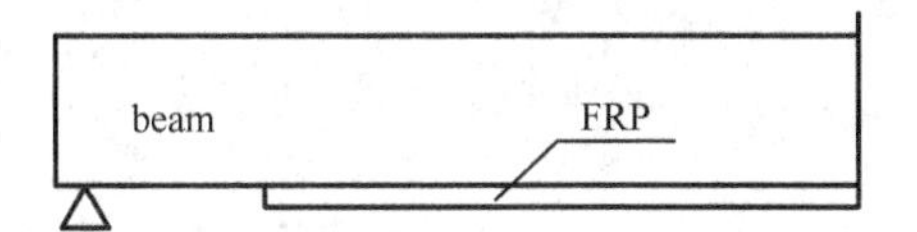

Fig.4-46　FRP composite panel model

The retrofit of bridges by using FRP includes the following stages:

(1) Removal of the concrete deck and the sidewalk.
(2) Steel repairs (replacing rusted rivets with bolts, fish-plating areas of section loss, etc.).
(3) Installation of temporary deck grating.
(4) Cleaning and painting steel superstructure.
(5) Removal of temporary grating and installation of light-weight FRP deck.
(6) Replacement of approach and bridge railings.

Load Testing

The methods to evaluate the bearing capacity of existing bridges, including static load test and dynamic load test, are introduced as follows.

1. Static load test method

At first, the static load test method selects the bridge hole with the most unfavorable calculation force and serious damage on the bridge structure, which is easy to set up the observation scaffold on, and then selects one or two sections controlled by the main internal forces. Then the external load (Fig.4-47), which is basically equivalent to the design load or the use of the load, is applied to test the strain, deflection deformation, crack and development of the control position and the control section of the bridge structure under load using the testing instrument (Fig.4-48). The transverse distribution coefficient and deformation of the pier and abutment are changed. The test results are compared with the calculated values of the structure theory and the allowable values of the code, to calculate the corresponding load levels and bearing capacity, or correct the structural resistance. If the test load to the maximum load is normal, the test value is consistent with the analytical value, and the deflection and crack are within the prescribed allowable range, which can be considered that the bridge can pass safely under this load.

Fig.4-47　static load test

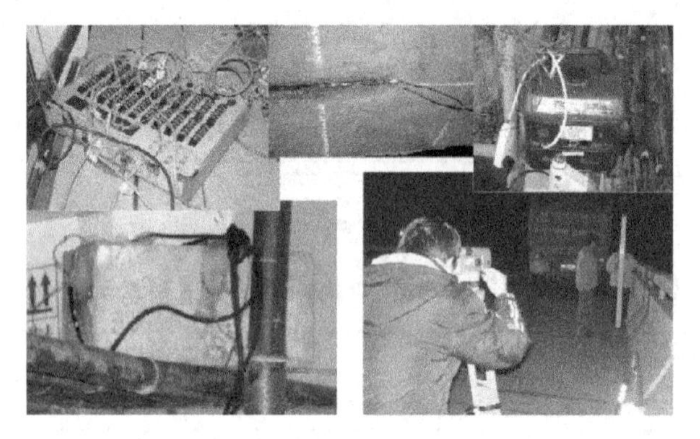

Fig.4-48　data collection

The static load test method is intuitionistic and reliable, but it can not replace the conventional evaluation method. It can only be regarded as a supplement to the means of obtaining information and the method of analysis.

2. Dynamic load test method

The dynamic load test is to test the frequency and amplitude of the control parts of the bridge structure by means of the exciting loads of the running car, the jumping car, the brake dynamic load or the steady state and transient state of the vehicle on the bridge(Fig.4-49). The dynamic parameters such as damping and impact coefficient are analyzed and compared with the corresponding calculated or empirical values to evaluate the bearing capacity(Fig.4-50).

Fig.4-49　jumping load

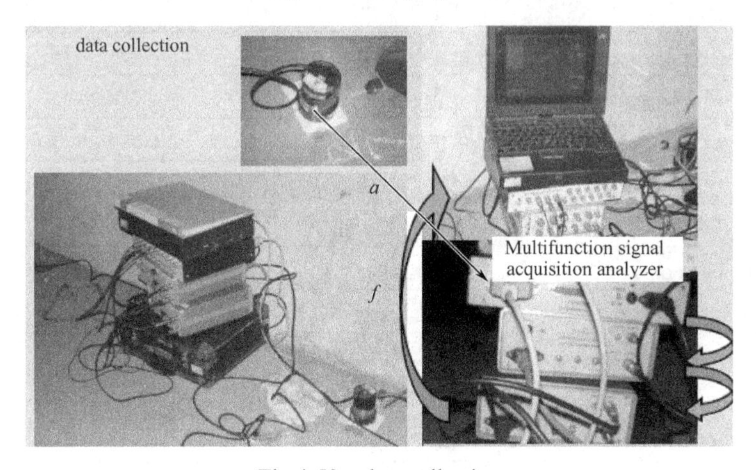

Fig.4-50　data collection

4.5.2　Notes

1．概念区分

旧桥：桥梁开始出现损坏。

危桥：桥梁损坏严重，影响其功能的发挥。

维修：旧桥正常的维护和修缮工作，使桥梁恢复原有性能、形状，修复排、防水设施，管理和养护桥面伸缩缝、围栏、支座（Retrofit，rehabilitation）。

加固：对旧、危桥提高载荷等级，**恢复**其原有的设计标准（Reinforce，consolidation）。

增强，补强：对旧危桥提高载荷等级，并**超过**其原有的设计标准（Strengthen）。

4.5.3　New Words and Expressions

deteriorate [dɪˈtɪəriəreɪt]	恶化，变坏

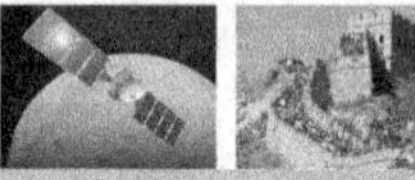

staggeringly [ˈstæɡərɪŋlɪ]	难以置信地；令人震惊地
replacement [rɪˈpleɪsmənt]	代替
deficient [dɪˈfɪʃnt]	不足的，缺乏的，有缺陷的
structural deficiency	结构性缺陷
unsafe [ʌnˈseɪf]	不安全的，危险的
collapse [kəˈlæps]	坍塌
functional deficiency	功能性缺陷
load rating	额定负载，额定载荷
weight restriction	限重
de-icing salt	除冰盐
the Ministry of Transport	交通运输部
obsolete [ˈɒbsəli:t]	旧的，过时的
obsolete bridge	旧桥
dangerous bridge	危桥
linear meter	延米
fiber reinforced polymer(FRP)	纤维增强复合材料
composite [ˈkɒmpəzɪt]	复合材料
alternative [ɔ:lˈtɜ:nətɪv]	可供选择的
community [kəˈmju:nəti]	社区，共同体
ratio [ˈreɪʃiəʊ]	率，比例，比值
deteriorated component	破损构件
load posting	交通限载
prolong [prəˈlɒŋ]	延长，拉长，拖延
service life	服务年限，使用年限
truss bridge	桁架桥梁
installation [ˌɪnstəˈleɪʃn]	安装
inconvenience [ˌɪnkənˈvi:niəns]	不方便，麻烦
rivet [ˈrɪvɪt]	铆钉
plate [pleɪt]	金属板，在……上覆盖金属板
grating [ˈɡreɪtɪŋ]	格栅、栅栏
railing [ˈreɪlɪŋ]	栏杆
load testing	载荷试验
static load test	静载荷试验
dynamic load test	动载荷试验
internal force	外载荷，外力
external force	内载荷，内力
design load	设计载荷
instrument [ˈɪnstrəmənt]	仪器

transverse [ˈtrænzvɜːs]	横向的
distribution [ˌdɪstrɪˈbjuːʃn]	分布
coefficient [ˌkəʊɪˈfɪʃnt]	系数；（测定某种质量或变化过程的）率；程度
calculated value	计算值
consistent [kənˈsɪstənt]	一致的；连续的
analytical value	分析值
intuitionistic [ɪntjuːɪʃəˈnɪstɪk]	直觉的，直观的
reliable [rɪˈlaɪəbl]	可靠的
supplement [ˈsʌplɪmənt]	增补，补充
data collection	数据收集
frequency [ˈfriːkwənsi]	频率
amplitude [ˈæmplɪtjuːd]	振幅
exciting load	激振载荷
brake [breɪk]	制动，刹车
transient[ˈtrænziənt]	短暂的，临时的
dynamic parameter	动态参数
damping [ˈdæmpɪŋ]	阻尼，减幅
impact coefficient	冲击系数

知识分布网络

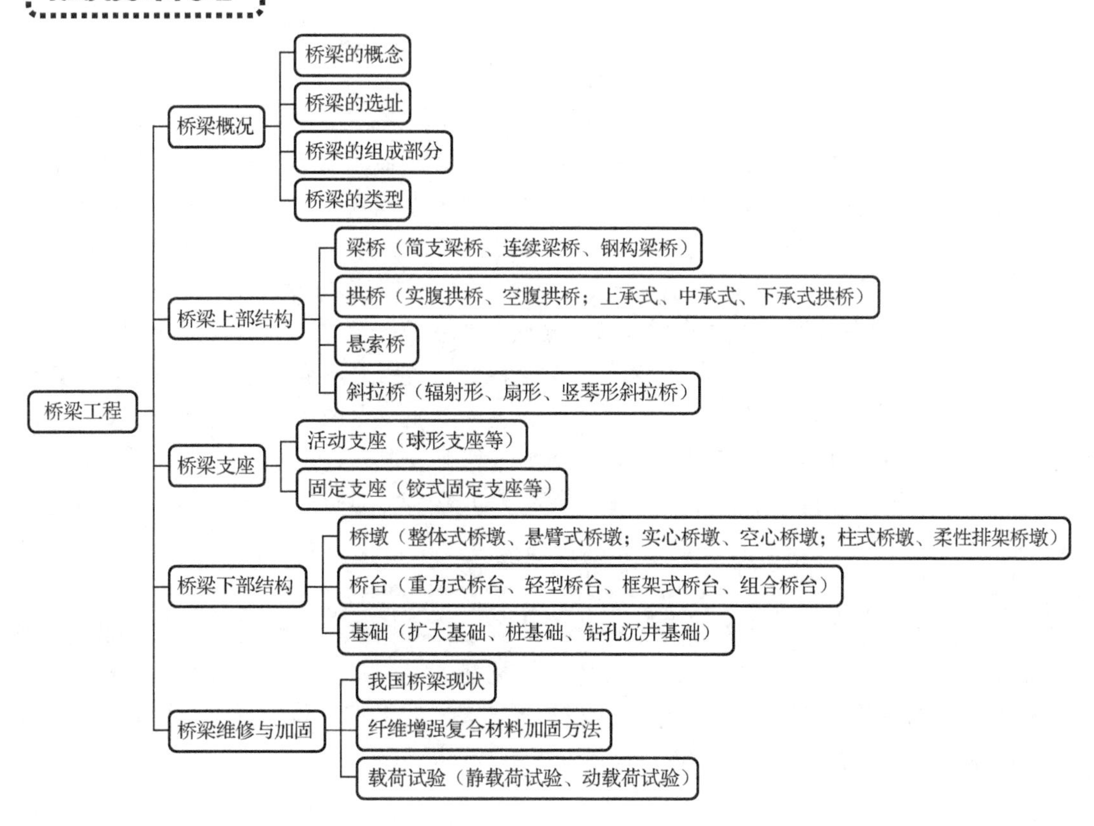

知识梳理与总结

桥梁工程是一门工程类应用科学，它是土木工程中结构工程的一个分支，它是包括设计理论与技术、施工技术与工艺、养护管理与技术等在内的一个完整体系。本章内容是该教材的重点部分，课文多取自近年英文原版书刊及网络，基本上包含了桥梁工程的常用专业英语词汇。取材时考虑难度适中，既注重对学生的英语语言的训练，又重视桥梁工程专业领域新知识的传递。每课之后都附有专业英语词汇表和注释，希望学生能够明确学习目标，依据教材、词典、多媒体课件、视频等学习手段和工具掌握桥梁工程的专业英语词汇，并且可以阅读简单的英语文献资料。本章必须掌握的知识要点有：

1. 在桥梁概述部分，需要掌握桥梁的概念、选址、基本组成部分及桥梁类型等知识点及专业英语词汇。

2. 桥梁结构由上部结构、支座和下部结构组成。桥梁上部直接承受活载荷作用的结构称为上部结构。桥梁上部结构由桥面构造和桥跨构造组成，每一种类型有其自身的结构特点和跨径。此教材对梁桥、拱桥、悬索桥、斜拉桥等不同桥的上部结构进行了阐述，要求学生掌握梁桥（简支梁桥、连续梁桥、刚构梁桥）、拱桥（实腹拱桥、空腹拱桥；上承式、中承式、下承式拱桥）、悬索桥、斜拉桥（辐射形、扇形、竖琴形斜拉桥）的概念及受力特点等内容及专业英语词汇。

3. 桥梁支座是桥梁结构中的一个重要组成部分，它位于桥梁上部结构和下部结构之间，其主要功能是将载荷从上部结构传递给下部结构，并适应活载荷等因素所产生的位移。本章对桥梁支座的类型、受力特点等进行了介绍。

4. 桥梁的下部结构类似于单层建筑的墙、柱和基础。在桥梁中，下部结构是上部结构的支撑结构，它由桥墩、桥台、翼墙和基础组成。单跨桥梁两端都有桥台，为桥梁提供竖向和横向支撑，并充当挡土墙，抵抗桥面填土的侧向压力。多跨桥梁需要桥墩支撑桥台以外的跨端。基础的类型取决于持力层的深度和安全承载能力，也受到不同类型桥面板差异沉降的限制。在一般情况下，基础分为三类：条形基础、桩基础和钻孔沉井基础。

5. 在桥梁维修与加固小节中介绍了我国桥梁的现状及纤维增强复合材料加固方法和载荷试验（静载荷试验、动载荷试验）。现代复合材料以 20 世纪 40 年代 CFRP（Carbon Fiber Reinforced Polymer/Plastic）的出现为标志，目前已研发出了具有各种优异性能的聚合物基复合材料，包括玻璃纤维、碳纤维、芳纶纤维等增强复合材料。在航空航天领域、现代国防工业中 FRP 首先得到应用。在民用工业如机械工业、交通运输、建筑工业以及生物医学、体育等领域，FRP 由于其优异性能也得到广泛应用。工程材料的进步及新材料的出现，历来是土木结构工程发展的动力。碳纤维材料的出现和成功应用使土木工程加固与补强技术研究更上一个台阶。碳纤维是一种新型建材，因其质量轻、耐腐蚀、片材很、抗拉强度高而被广泛应用。碳纤维布（片）加固法亦被视为梁桥加固补强、提高承载能力，尤其在高度受限制时的首选方法。其施工工艺也很简单，适用于钢筋混凝土受压柱，以提高延展性、耐久性的加固；亦可用于梁、板的加固。

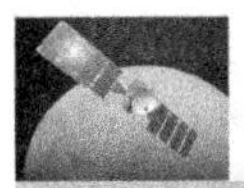

思考与练习题 4

扫一扫看第 4 章习题答案

Task 1: Match the words given in Column A with the meanings given in Column B.

A	B
1. truss bridge	a. an approximately triangular surface area between two adjacent arches
2. clearance	b. a low wall alone the edge of a bridge to stop people from falling
3. retaining wall	c. the middle support of a bridge to reduce the span
4. tie beam	d. the end support of a bridge superstructure
5. rivet	e. a bridge supported by trusses
6. spandrel	f. space and distance between two points
7. girder	g. similar
8. parapet	h. bricks or pieces of stone for building
9. abutment	i. a metal pin that is used to fasten two pieces of leather,metal together
10. pier	j.a wall that is built to prevent the earth behind it from moving
11. shear force	k. a long beam used for bridges
12. tensile force	l. it is the center to center distance between two supports of bridge
13. backfill	m. a horizontal beam to prevent two other structural members from spreading apart or separating
14. streamlined	n. to make a passage, hole as the result of movement, especially over a long period
15. well-defined	o. something has a shape that allows it to move quickly through air or water
16. feasibility	p.a stress that produces an elongation of an elastic physical body
17. scour	q. having a clear and distinct outline
18. analogy	r. the quality of being doable
19. masonry	s. to fill a hole with the material that has been dug out of it
20. span	t. a deformation of an object in which parallel planes remain parallel but are shifted in a direction parallel to themselves

Task 2: Choose one correct answer from A, B, C and D.

1. “cost-effective”means____________________.

(A) high efficient　　(B) low efficient　　(C) high cost　　(D) low cost

2. Broadly, a bridge can be divided into two major parts______________.

(A) pier and abutment　　(B) superstructure and substructure

(C) approach and foundation　　(D) beam and bearing

3. ______________are provided in between the two extreme supports of the bridge(abutments) and in the bed of the river to reduce the span and share the total load coming over the bridge.

(A) piers　　(B) abutments　　(C) foundations　　(D) approaches

Task 3: Find out the synonyms for the underlined section.

1. The Ministry of Transport estimates that repairing deficient or obsolete bridges will cost more than 20 billion RMB.

(A) sorely (B) absolutely (C) fashionable (D) dated

2. Piers are intermediate supports in a multi-span bridge system.

(A) end (B) middle (C) instant (D) important

3. Approaches may be in embankment or in cutting depending upon the design of the bridge.

(A) excavation (B) trench (C) removal (D) backfilling

4. The thickness of slab is quite considerable but uniform, thereby requiring simple shuttering.

(A) supporting (B) tensioning

(C) compression (D) form engineering

5. Bow string girder bridges are economical when sufficient head room is needed under a bridge.

(A) house (B) first (C) clearance (D) interval

6. These bridges are flexible and hence the vertical oscillations will be more than the other bridges.

(A) displacements (B) movements (C) vibrations (D) expanding

7. Wingwalls can be constructed of cast-in-place concrete and shall be designed as retaining walls.

(A) precast (B) placing concrete in site

(C) installation (D)curing

Task 4: Translate the following into Chinese.

1. Simultaneously, bridge infrastructure in China is deteriorating at an alarming rate. Due to the staggeringly high cost of repair and replacement, most transportation agencies are unable to cope with this trendency. By the end of 2006, nearly 6 282 bridges in China had been classified as deficient. Structural deficiency does not imply that a bridge is unsafe or likely to collapse. More than 3 000 bridges are classified as structurally or functionally deficient because of poor deck conditions, lack of load ratings or weight restrictions. Much of the deck deterioration can be attributed to heavy application of road de-icing salts during the snowy winters experienced in the nation.

2. The banks on both sides of the river should have firm soil and be straight and well-defined to increase the stability of the bridge and reduce the possibility of the erosion of the banks. The bridge site should be at a place where the river is narrow and the flow is streamlined without serious whirls and cross currents to reduce the length of the bridge and to guard against scour.

3. Suspension cable must be anchored at both ends of the bridge because all loads acting on the bridge will be converted into the tension on the main cable. The main cable passes through the towers that is attached to caissons or cofferdams and connected to the deck through suspenders or cables, and then anchored in the foundation .The main forces in a suspension bridge are tension on

the main cables and compression on the pillars.

4. Bearing is a main part in bridge structures, which is positioned between the bridge superstructure and substructure. The key functions are to transmit loads from the superstructure to the substructure and to accommodate relative movements. The movements in the bridge bearings include translations and rotations. Creep, shrinkage, and temperature effects are the most common causes of the translational movements, which can occur in both transverse and longitudinal directions. Traffic loading, construction tolerances, and uneven settlement of the foundation are the common reasons of the rotations.

5. Foundation types depend mainly on the depth and safe bearing capacity of the bearing stratum, also are restricted by differential settlement due to the type of bridge deck. Bridge foundations are generally split into three categories: spread footings, pile foundations and drilled caisson foundations.

Task 5: Point out the types of the follow bridges.

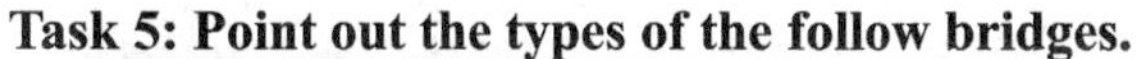

1.

____

2.

____

3.

____

4.

____

5.

____

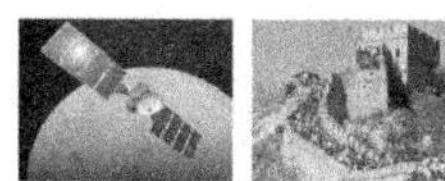

总词汇表

abrasive force	摩擦力
abutment [ə'bʌtmənt]	桥台
abutment cap	台帽
access design	路口设计
accessory [əkˈsesəri]	附属设施
accommodate [əˈkɒmədeɪt]	适应
acid [ˈæsɪd]	【化】酸；酸性物质
adjacent [əˈdʒeɪsnt]	邻近的，毗邻的
aerial photography	航空摄影
aerial survey camera	航空测量摄影仪
aerodynamics [ˌeərəʊdaɪˈnæmɪks]	空气动力学
aggregate	骨料；集料（可用于混凝土或修路等）；混凝料
aging	老化；老龄化；（酒等的）陈化；熟化
airplane	飞行器，飞机
alignment [əˈlaɪnmənt]	arrangement in a straight line 排成直线
alignment design	线形设计
alkali [ˈælkəlaɪ]	【化】碱
allowable stress intensities	允许应力强度
alloy [ˈælɔɪ]	合金
alternative [ɔ:lˈtɜ:nətɪv]	可供选择的
ambush[ˈæmbʊʃ]	伏击，埋伏
amplitude [ˈæmplɪtju:d]	振幅
analogous [əˈnæləgəs]	相似的
analytical value	分析值
anatomy [əˈnætəmi]	解剖，分解，分析；（详细的）剖析
anchor [ˈæŋkə]	n.锚；v.锚固于
anchor bolt	锚栓
anchor stud	锚固螺栓
anchorage [ˈæŋkərɪdʒ]	锚固装置
ancillary[ænˈsɪləri]	辅助的；补充的；附属的；附加的
applied load	外加负载，施加载荷
approach	小径，引路

approach roadway	引道，引桥
arc [ɑ:k]	弧，弧度
arch [ɑ:tʃ]	拱圈
arch axis	拱轴线
arch bridge	拱桥
arterial [ɑ:ˈtɪərɪəl]	主干道，干线（指主要公路、河流、铁路线等）
artificial aggregate	人工合成骨料
asphalt [ˈæsfælt]	沥青；柏油
asphalt binder	沥青结合料
asphalt concrete	沥青混凝土
asphaltic	柏油的
at grade	【美】在同一水平面上；
at ordinary temperature	在常温下
at right angles to	与……成直角
autobahn [ˈɒtəʊba:n]	（德国、奥地利或瑞士的）高速公路
automobile [ˈɔ:təməbi:l]	汽车
autoroute [ɔ:təʊ'ru:t]	（法国和法语地区的）高速公路
auxiliary lane	辅道
avalanche [ˈævəlɑ:nʃ]	雪崩
award	授予
axial force	轴向力
axial load	轴向载荷
base	基层
base course	基层
basement	【建】基底，底部
battered pile	斜桩
be responsible for	对……负责
beam	梁
beam bridge	梁桥
bearing [ˈbeərɪŋ]	支座
bearing capacity	承载力
bearing pad	承重垫片
bearing stratum	持力层
bending ['bendɪŋ]	弯矩

bending moment	弯矩
bent	排架
bentonitic	（含有）膨润土（皂土、斑脱土）的
bid	出价，投标
bid bond [bid bɔnd]	投标保证；履约担保书；投标保证金；押标金
bidder [ˈbɪdə(r)]	出价者，投标人
bidding document	招标文件
binding agent	黏合剂，结合剂
bitumen [ˈbɪtʃəmən]	沥青；柏油
bituminous [bɪˈtju:mɪnəs]	含沥青的
blast furnace	鼓风炉
bleeding	泛油
blind ditch	盲沟
bolt [bəʊlt]	螺栓
Bonn [bɔn, bɔ:n]	波恩（德国城市）
bored pile	钻孔灌注桩
borrow pit	取土坑
borrower	借款人
bottleneck	瓶颈
bottom-through arch bridge	下承式拱桥
box beam	箱梁
brake [breɪk]	制动，刹车
branch	分科
break-even [breik ˈi:vən]	收支平衡；不赔不赚；经济的
brick [brɪk]	砖
bridge engineering	桥梁工程
bridge site	桥址
brittle-fracture	脆裂
bubble [ˈbʌbl]	泡，水泡
buckling	弯折，压曲
bulge[bʌldʒ]	膨胀
bypass	（绕过城市的）旁路，旁道，支路
byproduct	副产品
cable [ˈkeɪbl]	缆、索

cable-stayed bridge	斜拉桥
caisson ['keɪsən]	沉箱
caisson cap	沉井帽
calcium aluminate	铝酸钙
calcium silicate	硅酸钙
calculated value	计算值
calculation	计算
California Bearing Ratio (CBR)	加州承载比
canal	运河，沟渠
cantilever [ˈkæntɪli:və]	（桥梁或其他构架的）悬臂，悬桁，伸臂
cantilever beam bridge	悬臂梁桥
cantilevered pier	悬臂式桥墩
canyon [ˈkænjən]	峡谷
carbon [ˈkɑ:bən]	【化】碳
carbon disulfide	二硫化碳
carriageway	行车道
cast	铸型，铸造
cast in place	浇筑，现场浇筑
cast-in-place concrete	现浇混凝土
catenary arch bridge	悬链拱桥
cement [sɪˈment]	水泥，胶合剂
center line	中心线
central median	中央分隔带
centre line	中心线
Cessna cessna aircraft company	塞斯纳飞机公司
channel	沟渠
channelization	渠化
Chinese Labour Law	中国劳动法
circle arch bridge	圆弧拱桥
circular curve	圆曲线
civil engineering	土木工程
civil works	土木工程，土建
clarification [ˌklærəfɪ'keɪʃn]	说明
clay	黏性土

cleanse [klenz]	净化；使……清洁；清洗
clearance	净空
clearance[ˈklɪərəns]	净空，空隙
cloverleaf interchange	苜蓿叶形立体交叉道
coal tar	煤焦油
coalesce [ˌkəʊəˈles]	联合，合并
coarse	粗的
coarse aggregate	粗集料
coat	涂料层；覆盖层
cobble	卵石
code	规范
coefficient [ˌkəʊɪˈfɪʃnt]	系数；（测定某种质量或变化过程的）率；程度
cofferdam ['kɒfədæm]	沉箱，围堰
cohesion [kəʊˈhiːʒn]	凝聚，内聚；（各部的）结合；【力】内聚力
cohesive	黏性的，黏质的，有黏着力的
collapse [kəˈlæps]	坍塌
collector lane	辅道
Cologne [kəˈləʊn]	科隆（德国城市）
column [ˈkɒləm]	柱，梁
combination of bids	组标
combined highway and railway bridge	公铁两用桥
combined system bridge	组合体系桥
community [kəˈmjuːnəti]	社区，共同体
compact	v.压实
compaction	n.压实
compaction device	压实设备
compactor	【美】垃圾捣碎机，压土机，夯土机；夯具
competent rock	强岩层
composite material	合成材料，复合材料
compress [kəmˈpres]	压紧，压缩
compression [kəm'preʃn]	压力
compressive load	压力载荷
computer graphics [ˈgræfɪks]	计算机图形学
computer monitor	计算机显示器

computer technique	计算机技术
concave [kɒnˈkeɪv]	凹面的，凹面
concrete	混凝土
configuration [kənˌfɪgəˈreɪʃn]	构造，外形，布局
confine [kənˈfaɪn]	限制；局限于；禁闭；管制；界限，范围
conical slope	锥形护坡
connectivity	连接性
consistent [kənˈsɪstənt]	一致的；连续的
construction	施工
construction accidents	施工事故
construction tolerance	施工误差
consultant[kənˈsʌltənt]	顾问
content	内容，目录；容量，含量
continuous beam bridge	连续梁桥
contour [ˈkɒntʊə(r)]	（地图上表示相同海拔点的）等高线
contract [ˈkɒntrækt]	合同
contractor	承包人
control point	控制点，检测点
controlled-access highway	限制进入的道路，快速道
convert [kənˈvɜ:t]	转变，转化
convex [ˈkɒnveks]	凸面的
coordinate [kəʊ'ɔ:dɪneɪt]	【数】坐标
cost-effective	有成本效益的，划算的；合算的
counterfort [ˈkaʊntəfɔ:t]	扶壁
crack [kræk]	断裂，开裂，裂缝
cracking	破裂，裂缝
crane safety	起重机安全
creep [kri:p]	徐变
crest vertical curve	凸形竖曲线
criteria[kraɪˈtɪərɪə]	标准，准则（criterion 的名词复数）
critical length of grade	边沟的临界长度
crocodile cracking	龟裂
cross hair	交叉瞄准线，（光）标线，十字线
cross section	横断面，横截面

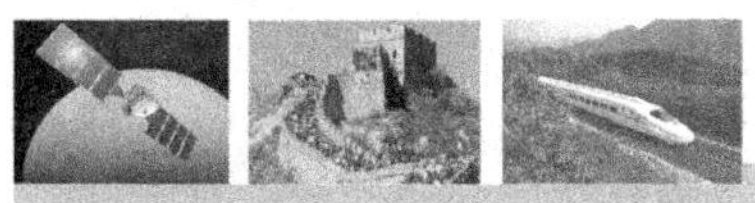

cross section design	横断面设计
crossing	交叉路口，十字路口
crude oil [kru:d ɔil]	原油
crushed stone	碎石
cubic centimeter	立方厘米
culvert[ˈkʌlvət]	涵洞
cured material	固化材料
curvature [ˈkɜ:vətʃə(r)]	曲率
curve length	曲线长度
curved arch bridge	双曲拱桥
curved bearing	曲面支座，球形支座
cut	挖方
cut-back asphalt	【化】稀释沥青；油溶沥青；轻制沥青
cycle track	自行车车道
damping ['dæmpɪŋ]	阻尼，减幅
dangerous bridge	危桥
data collection	数据收集
dead load	永久固定的载荷（如房屋、桥梁等）；自重；恒载荷
dead weight	静负载，固定负载
debris [ˈdebri:]	碎片，残骸
deck [dek]	桥面
deck arch bridge	上承式拱桥
deck bridge	上承式桥
deck system	桥面系统
deep foundaiton	深基础
deficient [dɪˈfɪʃnt]	不足的，缺乏的，有缺陷的
deflection [dɪˈflekʃn]	偏斜，偏移，歪曲；偏斜度；挠度
deflection angle	偏转角，偏角；
deform	使变形
deformation	变形
degree of curvature	弯曲度，曲度
de-icing salt	除冰盐
demerge [ˌdi:ˈmɜ:dʒ]	拆分，分离
densification [densɪfɪ'keɪʃən]	增浓作用，稠化（作用）；致密；捣实；夯实

density	密度
dependable	可靠的
depict	描述
derivative [dɪˈrɪvətɪv]	衍生物，派生物
design	设计
design load	设计载荷
designed flood level	设计水位
destructive distillation	分解蒸馏，干馏
deteriorate [dɪˈtɪəriəreɪt]	恶化，变坏
deteriorated component	破损构件
deterioration [dɪˌtɪərɪəˈreɪʃn]	恶化；变坏；退化
detrimental [ˌdetrɪˈmentl]	有害的（人或物）；不利的（人或物）
diameter[daɪˈæmɪtə]	直径
diamond interchange	钻石形立体交叉道
differentiate [ˌdɪfəˈrenʃieɪt]	区分，区别，辨别
digital [ˈdɪdʒɪtl]	数字的
digital level	数字水准仪
digital mapping	数字制图
dimension	n.尺寸；【数】维；v.标出尺寸
directional ramp	直接式匝道
discharge	排放
disk bearing	板式支座
displacement	位移
dispose	处置，处理
disrupt	v.使中断；adj.中断的
dissolve [dɪˈzɒlv]	（使）溶解
distress	悲痛；危难，不幸
distribute	分布
distribution [ˌdɪstrɪˈbjuːʃn]	分布
distributor lane	辅道
ditch	沟渠
domestic refuse	生活垃圾
dowel	木钉，销子
dowel bar	（混凝土路面）销钉，接缝条；传力杆

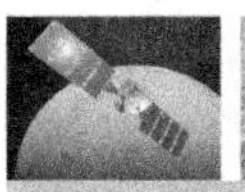

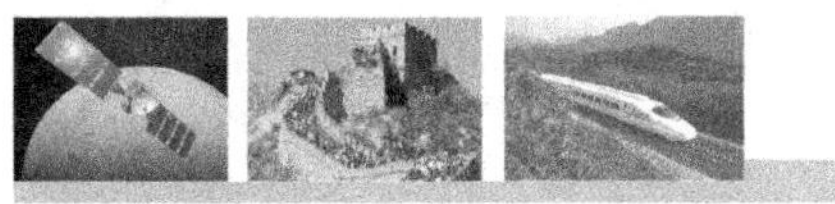

downward	向下的
drainage [ˈdreɪnɪdʒ]	排水，排水系统
drainage ditch	排水沟
drainage system	排水系统
drainage tube	排水管
drainage[ˈdreɪnɪdʒ]	排水系统
draw [drɔ:]	拉，拔出
drawing	图样，图纸
drilled caisson foundation	钻孔沉箱基础
driven cast in-situ pile	沉管灌注桩
driven pile	灌注桩
ductility [dʌk'tɪlɪtɪ]	延展性，柔软性
durability [ˌdjʊərə'bɪlətɪ]	耐久性；持久性
dynamic load test	动载荷试验
dynamic parameter	动态参数
earthen fill	填土
earthquake belt	地震带
earthquake load	地震载荷
earthwork [ˈɜ:θwɜ:k]	土方（工程）
eject[iˈdʒekt]	喷出；驱逐；强制离开
elastic design	弹性设计
elastic-plastic displacement	弹塑性位移
elastomer	橡胶
elastomeric bearing	橡胶支座
elastomeric disk	橡胶板
elastomeric plate	橡胶板
electrical safety	电气安全
electronic [ɪˌlekˈtrɒnɪk]	电子的
electronic distance measurement	电子测距仪
electronic equipment	电子设备
electronics [ɪˌlekˈtrɒnɪks]	电子学
element	元素
elevation [ˌelɪˈveɪʃn]	高度，海拔
eliminate	消除

embankment [ɪmˈbæŋkmənt]	路堤
embed [ɪm'bed]	把……嵌入
embed[ɪm'bed]	把……嵌入
emergency aid procedure	紧急措施
emulsified asphalt	乳化沥青
emulsify [ɪˈmʌlsɪfaɪ]	使乳化
end bearing pile	端承桩
engineering project	工程项目
enhance [ɪnˈhɑ:ns]	提高，增加；加强
entail[ɪnˈteɪl]	牵涉；需要
entrance ramp	入口匝道
epoxy[ɪˈpɒksi]	环氧的；环氧树脂
equivalent[ɪˈkwɪvələnt]	相等的，相当的
erection [ɪˈrekʃn]	安装，装配，建设
erosion [ɪ'rəʊʒn]	腐蚀，侵蚀
evaluation [ɪˌvælju'eɪʃn]	评价
examination[ɪgˌzæmɪˈneɪʃn]	审查，检查
excavate[ˈekskəveɪt]	挖凿，开挖
excavation [ˌekskəˈveɪʃn]	发掘，挖掘
excess [ɪkˈses]	超过（的）；超额量（的）
excess load	超载荷
exciting load	激振载荷
exit ramp	出口匝道
expand [ɪkˈspænd]	扩展；发展；张开；展开，扩大
expansion bearing	可动支座，活动支座
expansion joint	伸缩接头，伸缩缝
expel	排出
expiration [ˌekspəˈreɪʃn]	截止日期
expressway	高速公路
extension [ɪkˈstenʃn]	延期
extent	程度
external [ɪkˈstɜ:nl]	外面（的），外部（的）
external force	内载荷，内力
facilitate [fəˈsɪlɪteɪt]	有助于，促进，助长

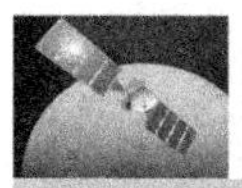

facility	设施，设备
falling weight deflectometer	落锤式弯沉仪
fan design	扇形
fatigue [fəˈti:g]	疲劳
feasible[ˈfi:zəbl]	可行的；行得通的
fiber reinforced polymer(FRP)	纤维增强复合材料
fill	填方，填
filled spandrel arch bridge	实腹拱桥
film	薄层；薄膜
fine	细的
fine aggregate	细集料
fire protection and prevention	防火
first aid procedure	急救措施
fixed bearing	固定支座
flag	薄层，薄层砂岩
flexible pavement	柔性路面
flyover	立交桥，高架公路
flyover ramp	立交桥匝道
footbridge	人行桥
footpath	人行道
forfeit [ˈfɔ:fɪt]	（因违反协议、犯规、受罚等）丧失，失去
formation	形成；构成，结构
foundation	基础
four-way interchange	四方向立体交叉道
fracture[ˈfræktʃə]	（使）折断，破碎
framing configuration	框架结构
free board	安全高度
freeway	高速公路
freeway junction	高速公路立体交叉路口
freeze	冻结；严寒时期
frequency [ˈfri:kwənsi]	频率
frictional ['frɪkʃənəl]	摩擦的，摩擦力的
frictional pile	摩擦桩
frost	霜冻

frost heave	冻胀，冰冻膨胀
full-depth asphalt pavement	全厚式沥青路面
function [ˈfʌŋkʃn]	功能，作用
functional deficiency	功能性缺陷
geometry[dʒiˈɒmətri]	几何学
girder [ˈgɜːdə]	大梁，主梁
girder bridge	箱梁桥
Global Positioning Satellite systems	GPS，全球定位卫星系统
gradation	级配
grade	坡度
grade intersection	平面交叉
grade line	纵坡线，坡度线
grade separation	立体交叉
graded aggregate	级配骨料，分级粒料
gradeline elevation	纵坡线的高程
graduated arc [ˈgrædjʊeɪtɪd ɑːk]	分度弧
graduated circular	分度圆
grain size	粒径
grand bridge	特大桥
grant[grɑːnt]	承认；同意；准许；授予
granular	粒状的，颗粒的
grating [ˈgreɪtɪŋ]	格栅、栅栏
gravel	砾石
gravity abutment	重力式桥台
grid roller	网格压印路碾，网格压路机
ground anchorage	地锚结构
ground surevey map	勘测图
ground survey	地面测量
grout [graʊt]	薄泥浆，水泥浆
guard stone	缘石
guardrail [ˈgɑːdreɪl]	护栏
half-through arch bridge	中承式拱桥
hammerhead pier	锤头式桥墩
handrail	扶手，栏杆

hard shoulder	硬路肩
harp design	竖琴形
haunch [hɔ:ntʃ]	n.腰腿（肉）；v.加腋（加腋是指在整体结构的转角处同时加大两个相交截面的面积，一般做成三角形，与相交的结构同时浇筑。加腋部分应适当配置构造钢筋。）
head room	净空高度
heavy traffic	大流量交通
heavy-duty	重型的，重载的
heel [hi:l]	脚后跟
hexagonal[heks'ægənl]	六角形的，六边形的
high flood level	高水位
high-quality	高质量
highway	公路，（尤指城镇间的）公路，干道，交通要道
highway alignment design	公路线型设计
highway bridge	公路桥
highway engineering	公路工程（学）
highway geometric design	公路几何设计
highway interchange	公路立体交叉道
highway location	公路定线
highway ramp	公路匝道
hinge	铰链，合页；关键，转折点
hollow beam	空心梁
hollow pier	空心式桥墩
horizontal [ˌhɒrɪˈzɒntl]	adj.水平的；地平线的　n.水平线；水平面
horizontal alignment	水平定线，水平线型
horizontal axis	水平轴
horizontal component force	水平分力
horizontal curve	平面曲线，水平曲线
horizontal/vertical angle	水平角/垂直角
hostile[ˈhɒstaɪl]	敌人的，敌方的
hydraulics [haɪˈdrɔ:lɪks]	水力学
hydrocarbon [ˌhaɪdrəˈkɑ:bən]	【化】碳氢化合物，烃
illustration	图示，图解
impact coefficient	冲击系数
impact load	冲击负载

in accordance with	与……一致
in close conformity to	与……密切一致
in close proximity to	与……靠得很近
in the preliminary phase of project	工程初期
inconvenience [ˌɪnkənˈviːniəns]	不方便，麻烦
inert	【化】惰性的
infrastructure	基础设施；基础建设
insert [ɪnˈsɜːt]	插入，嵌入
in-situ	原位，现场
in-situ moisture content	原位含水量
inspection	检查，视察；检验，审视
installation [ˌɪnstəˈleɪʃn]	安装
instigate [ˈɪnstɪgeɪt]	教唆；煽动；激起
instrument [ˈɪnstrəmənt]	仪器
interchange	互通式立体交叉道
interlock	互锁，嵌锁
internal force	外载荷，外力
internal gross forces	总内力
international bidding	国际投标
intersection	相交，
interval [ˈɪntəvl]	间隔；（数学）区间
intervention	介入，干涉，干预；调解，排解
intuitionistic [ɪntjuːɪʃə'nɪstɪk]	直觉的，直观的
jobsite safety policy	现场安全策略
joint [dʒɔɪnt]	关节；接合处
junction[ˈdʒʌŋkʃn]	（公路的）交叉路口
landscaping facility	绿化景观设施
lane	行车道
large span length	大跨度
lateral [ˈlætərəl]	侧面的，横向的
lateral load	横向载荷
latitude [ˈlætɪtjuːd]	维度
layout	布局
leak ditch	渗沟

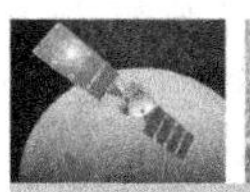

leeward [ˈliːwəd]	adj.& adv.背风的（地），下风的（地）；n.下风
left-bound highway	左行公路
length of the bridge	桥长
lens	透镜，镜头
level	水平仪，水准仪
level bar	水平杆，水平尺
level tube	水准测管，水准器，水准仪管
leveling head	校平头
leveling screw	校平【水准】螺旋；准平螺钉
life expectancy	使用期限
light beam	光束
light post	灯杆
light rail	轻轨
lighting facility	照明设施
lighting fixture	照明器材
lime [laɪm]	石灰
lime-bound concrete	石灰混凝土结合料
limestone quarry	石灰石采矿场
limit state design	极限状态设计
limited access road	限制进入的道路，快速道
linear [ˈlɪniə(r)]	线性的，直线的
linear meter	延米
linearity [ˌlɪnɪ'ærətɪ]	线性，直线性
liquefy [ˈlɪkwɪfaɪ]	（使）液化，溶解
liquid limit	液限
live load	活载荷
load	载荷
load bearing	承载，承重
load posting	交通限载
load rating	额定负载，额定载荷
load testing	载荷试验
load-bearing strength	承载强度
location design	定线设计
location technique	定位技术 the act of finding the position

log [lɒg]	原木
longitude [ˈlɒŋgɪtju:d]	经度
longitudinal	经度的；纵向的
longitudinal direction	竖向
long-term	长期
loop ramp	环形匝道
low water level	低水位
low-power laser beam	低能量激光束
lubricant [ˈlu:brɪkənt]	润滑剂，润滑油
lump sum	总价格，总金额，包干价
macro-level	宏观层面
magnitude [ˈmægnɪtju:d]	巨大，广大；重大，重要；量级；（地震）级数
main cable	主缆
main span	主跨
maintain	维护，保养
maintenance	维持，保持；保养，保管；维护；维修
major road	主干道，主要道路
manufacturer [ˌmænjuˈfæktʃərə]	制造商
marking	标记
marking	标记，记号
masonry [ˈmeɪsənri]	（筑墙或盖楼用的）砖石；石工工程，砖瓦工工程
masonry bridge	圬工桥，砌体桥
masonry plate	座板，下支座板
mass	质量
mathematical modeling	数学模型
maximum grade	最大纵坡
member	构件
merge	（使）混合；相融；融入
mineral aggregate	矿料
minimize	最小化
minimum ditch grade	最小边沟纵坡
minimum grade	最小纵坡
mitigate	减轻
moderate [ˈmɒdərət]	有节制的；稳健的，温和的；适度的，中等的

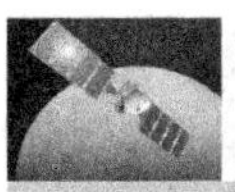

modifier	改性剂
moisture	水分，湿气
moisture content	含水量
monolithic pier	整体式桥墩
mortar [ˈmɔ:tə(r)]	灰浆；砂浆；胶泥
motorway	高速公路；汽车道；快车道
motorway junction	高速公路交叉路口
motorway service area	高速公路服务区
mount	安装
multi-level	多层的
multiple [ˈmʌltɪpl]	多重的；多个的
multi-span bridge	多跨桥梁
namesake	同姓名的人；同名的事物
natural aggregate	天然骨料，天然集料
nature	性质
noise barrier	噪声屏障
non-directional ramp	非直接式匝道
nonerodable	不可蚀的
non-freeway	非高速公路
non-linear	非线性
obsolete [ˈɒbsəli:t]	旧的，过时的
obsolete bridge	旧桥
obstacle [ˈɒbstəkl]	障碍（物）
obstruction	障碍物
octagonal[ɒk'tægənl]	八角形的，八边形的
off-ramp	出口匝道
on-ramp	入口匝道
on-site angle alignment	现场角度核准
open spandrel arch bridge	空腹拱桥
optical [ˈɒptɪkl]	视觉的，光学的
optimal	最佳的，最优的；最理想的
optimum [ˈɒptɪməm]	最适宜的，最佳的
orbiting satellite	轨道运行卫星
ordinary flood level	一般洪水位

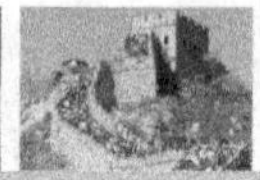

ordinary Portland cement (OPC)	普通硅酸盐水泥
overloading	超载
oxidation[ˌɒksɪ'deɪʃn]	氧化
panorama [ˌpænəˈrɑ:mə]	全景画；全景照片
parabolic arch bridge	抛物线拱桥
parabolic curve	抛物曲线
parabolic[ˌpærə'bɒlɪk]	抛物线的
parallel [ˈpærəlel]	平行的
parameter [pəˈræmɪtə]	【数】参数
parent material	原材料，母料
parkway	a wide road with trees，（有草木的）大路
parapet wall	护墙，矮墙
patch	补丁，修补
pavement	路面
pebble	卵石
pedestrian	行人
penetrate	穿透，刺入；渗入
penetration	渗透
penultimate [penˈʌltɪmət]	倒数第二（的）
periodic maintenance	定期养护
permanent bridge	永久桥
permanent load	永久负载，恒载荷
permanent steel casing	永久钢护筒
permeable	可渗透的
personal safety equipment	个人安全设备
petroleum [pəˈtrəʊliəm]	石油
petroleum asphalt	石油沥青
phase [feɪz]	【物理学】相位；
phenomenon [fəˈnɒmɪnən]	现象
photogrammetry[fəʊtə'græmətrɪ]	摄影测量法；摄影测量学
pier [pɪə]	桥墩
pier cap	墩帽
piled foundation	桩基础
pillar [ˈpɪlə]	柱，梁，墩

pin [pɪn]	销钉，铰
pin bearing	铰支座
pipeline	管道；输油管道
piston [ˈpɪstən]	【机】活塞
plain concrete	素混凝土
plane	（geometry 几何）flat or level surface 平面
planning	规划
plastic	塑性的
plastic limit	塑限
plasticity	塑性
plate [pleɪt]	金属板，在……上覆盖金属板
plate beam	板梁
plate compactor	板式压路机
pneumatic tire	充气轮胎
pneumatic-tired roller	胶轮压路机
polymer [ˈpɒlɪmə(r)]	多聚物；【高分子】聚合物
porous[ˈpɔ:rəs]	多空隙的，多孔的，能穿透的，能渗透的
Portland cement	【交】硅酸盐水泥
Portland cement concrete	普通硅酸盐混凝土
post-tensioning	后拉，后张
pot bearing	盆式支座
pothole [ˈpɒthəʊl]	（路面的）坑槽
pour	倾泻，倾倒
pozzolana [ˌpɒtswəˈlɑ:nɑ:]	火山灰（可用作水泥原料）
precast [ˌpri:ˈkɑ:st]	预浇铸的，预制的
precast concrete block	预制混凝土试块
prescribe [prɪˈskraɪb]	规定，指定
preservation [ˌprezəˈveɪʃn]	保存，保留；保护；防腐；维护，保持
preserve	保护；保持，保存
prestress ['pri:'stres]	给……预加应力
prestressed concrete	预应力混凝土
prestressed concrete bridge	预应力混凝土桥
pre-tensioning	预张法
processed aggregate	加工骨料，加工集料

profile [ˈprəʊfaɪl]	剖面
project planning	项目规划
prolong [prəˈlɒŋ]	延长，拉长，拖延
promptly [ˈprɒmptli]	迅速地；立即地
protection work	防护工程
prototype	原型
protruding [prə'tru:diŋ]	突出的、伸出的
prune [pru:n]	修剪（树木等）
PTFE(polytetrafluoroethylene)	聚四氟乙烯
pump	泵；用泵输送
pylon [ˈpaɪlən]	（架高压输电线的）电缆塔
quadrant	象限
RC bridge	钢筋混凝土桥
radial design	辐射形
radii['reɪdɪaɪ]	半径；半径（距离）（radius 的名词复数）
radius[ˈreɪdiəs]	半径
railing [ˈreɪlɪŋ]	栏杆
railway bridge	铁路桥
rain pipe	排（雨）水管
ramp [ræmp]	匝道
ramp metering	匝道车流调节
ratio [ˈreɪʃiəʊ]	比率，比例，比值
ratio of rise to span	矢跨比
react	（使发生）相互作用；（使起）化学反应
reaction [riˈækʃn]	反力，反作用力
readout	读出器，读出
ream [ri:m]	扩展，扩张
rebar	钢筋，螺纹钢筋
reconstruction	重建；再现；重建物；复原物
rectangular [rek'tæŋgjələ]	矩形的
rectify [ˈrektɪfaɪ]	改正，矫正
refine [rɪˈfaɪn]	提炼；精炼
reflector	反射器
refurbish[ˌri:ˈfɜ:bɪʃ]	刷新；使重新干净

regulation	管理；规章，规则，章程
rehabilitation	修复
reinforce	加固，加强
reinforced concrete (RC)	钢筋混凝土
reinforced concrete arch bridge	钢筋混凝土拱桥
reinforcement	加固，加强
reliable [rɪˈlaɪəbl]	可靠的
remote sensing	遥感技术
renewal[rɪˈnjuːəl]	重建，重生；更新，革新
renovation [ˌrenəˈveɪʃn]	翻新，修复，整修
repair	修理；纠正；恢复；弥补
replacement [rɪˈpleɪsmənt]	代替
reporting	报告
requirement	要求，规范
residential	住宅的
resilient modulus (MR)	回弹模量
respirator ['respəreɪtə(r)]	防毒面具，口罩
respiratory protection	呼吸防护
respiratory[rəˈspɪrətri]	呼吸的
restore	归还；交还；使恢复；修复
retain [rɪˈteɪn]	保持
retaining wall	挡土墙
retrofit [ˈretrəʊfɪt]	翻新，改型
ribbed arch bridge	肋拱桥
ribbed beam	肋梁
right angle	直角
rightmost	最右边
right-of-way	公路用地
right-turning	右转
rigid framed bridge	刚构桥
rigid pavement	刚性路面
rigid[ˈrɪdʒɪd]	刚性的
rigidity	刚度
river bed	河床

rivet [ˈrɪvɪt]	铆钉
road construction	筑路，道路建筑（施工，工程）
road intersection	道路交叉口
road marking	路标
road surface	路面
roadbed	路基，路床
roadside	路边，路旁
roadway	道路，车道
rock asphalt	岩沥青
rock dust	岩粉
rocker bearing	摇轴支座
roller	辊轮；滚筒，路碾，压路机
roller bearing	辊轴支座
rotary[ˈrəʊtəri]	旋转的；环行交叉路
rotation [rəʊˈteɪʃn]	转动
roundabout	绕道；环形交通枢纽
roundabout interchange	环岛形立体交叉道
route [ruːt]	路，路线
routine maintenance	日常养护
rubble [ˈrʌbl]	块石，瓦砾
rural roads	乡村道路
rutting	车辙
safety barrier	安全屏障
safety officer	安全主任
safety organization	安全工作人事结构
safety regulations	安全制度
sag vertical curve	凹形竖曲线
salt	【化】盐
scaffold [ˈskæfəʊld]	脚手架
scour [ˈskaʊə(r)]	冲刷
screen	筛分，筛选
sealing ring	密封圈
section	部分，断片，截面
sectional shape	截面形状

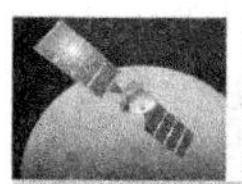

segment['segmənt]	部分、段落；分割
seismic ['saɪzmɪk]	地震的，由地震引起的
self weight of the structure	结构自重
semi-directional ramp	半直接式匝道
semi-solid	半固体，半固态的
semi-through bridge	中承式桥
separate contract	分项合同
service life	服务年限，使用年限，使用寿命
service loading	使用载荷，工作载荷
serviceability [sɜ:vɪsə'bɪlɪtɪ]	有用性，适用性；可维护性
settlement	沉淀，沉降
severe	严重的
shale	页岩
shear [ʃɪə(r)]	剪力
shear failure	剪力破坏
sheep-foot roller	羊足路碾，羊足压路机
shoulder	路肩
shoulder slope	路肩横坡
shrinkage ['ʃrɪŋkɪdʒ]	收缩
side belt	路缘带
side ditch	边沟
side slope	边坡
sidestep	侧向台阶
sidewalk	人行道
sight distance	视距
signal	信号
signpost	指示牌，标志杆；路标
silt	粉质土
simply supported beam bridge	简支梁桥
simultaneous [ˌsɪməl'teɪnɪəs]	同时的，一致的
simultaneously[ˌsɪməl'teɪnɪəslɪ]	同时地
single storey building	单层建筑物
single-span bridge	单跨桥梁
situate ['sɪtʃueɪt]	使位于，使处于……地位（位置）

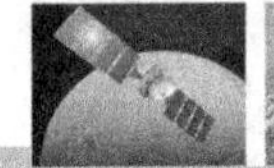

skeleton [ˈskelɪtn]	（建筑物等的）骨架
skid-resistant	防滑的
skyscraper [ˈskaɪˌskrepɚ]	摩天大楼，超高层大楼
slab [slæb]	平板，混凝土路面
slab bridge	板桥
slag	矿渣，熔渣
slender [ˈslendə]	苗条的，薄弱的
slide [slaɪd]	滑落
slip road	高速公路会交点
slope [sləʊp]	斜坡；斜面；倾斜；斜率
slot	窄缝
soft coal	烟煤
soil grain	土粒
sole plate	底板，上支座板
solid beam	实心
solid pier	实心式桥墩
soluble [ˈsɒljəbl]	【化】可溶的
solvent [ˈsɒlvənt]	【化】溶剂
span	跨径
span arrangement	跨径布置
spandrel ['spændrəl]	拱肩
spandrel structure	拱上结构，拱上建筑
specification [ˌspesɪfɪˈkeɪʃn]	规范
speed limit	速度限制
spherical bearing	球形支座
spindle [ˈspɪndl]	轴
spiral[ˈspaɪrəl]	螺旋形的；盘旋的
spirit level	（气泡）水准仪
spray	喷洒，喷射
spread footing	扩大基础
squeeze [skwi:z]	挤压，施加压力
stability [stəˈbɪləti]	稳定（性）
stabilization [ˌsteɪbəlaɪ'zeɪʃn]	稳定性；稳定化
stack interchange	环状形立体交叉道

stadium [ˈsteɪdiəm]	运动场；体育场
staggeringly [ˈstæɡərɪŋlɪ]	难以置信地；令人震惊地
stake holder	股份持有人
state of equilibrium	静力平衡
static load test	静载荷试验
steel arch bridge	钢拱桥
steel bridge	钢桥
steel reinforcing bar	钢筋条
steel reinforcing rod	钢筋杆
steel tube	钢管
steel-wheeled roller	钢轮压路机
stem [stem]	墩身，台身
stereo-plotter	立体绘图仪
stiff [stɪf]	坚硬，刚度
stiffness [stɪfnəs]	刚度
stipulate [ˈstɪpjuleɪt]	（尤指在协议或建议中）规定，约定，讲明（条件等）
stockpile [ˈstɒkpaɪl]	n.（原料，食品等的）储备，备用物资；v.大量储备
stone arch bridge	石拱桥
straight grade line	直坡线
strain[streɪn]	应变
strata [ˈstrɑːtə]	地层；岩层（stratum 的名词复数）
streamlined ['striːmlaɪnd]	（指汽车、飞机等）流线型的
strength	强度
strengthen	加强，巩固
stress	应力
stress intensity	应力强度
structural analysis	结构分析
structural deficiency	结构性缺陷
structural design	结构设计
structural element	结构构件
structural engineering	结构工程
structural system	结构体系
structure failure	结构失效破坏
subbase	底基层

subgrade	路基
subgrade strength	路基强度
submit [səbˈmɪt]	呈报，提交
subsidence	下沉，沉降
subsoil [ˈsʌbsɔɪl]	地基土
substructure	下部结构
subtraction	减法
suction [ˈsʌkʃn]	吸，抽吸
sufficient	足够的，充足的
suitability [ˌsjuːtəˈbɪlətɪ]	合适，适合
sulphur extended asphalt	掺硫沥青
superelevation	（公路的）超高（公路转弯处外侧比内侧高出的程度）
superimposed load	附加载荷
superior [suːˈpɪəriə]	（级别）较高的；（质量）较好的；（数量）较多的
superstructure	上部结构
supplement [ˈsʌplɪmənt]	增补，补充
supporting force	支承力
suppress [səˈpres]	镇压，压制；禁止（发表）；阻止……的生长（或发展）
surface abrasion	路面磨损
surplus	过剩的；多余的
survey [ˈsɜːveɪ]	测量；勘测；测绘
susceptibility	敏感性
suspend [səˈspend]	悬挂
suspender [səˈspɛndɚ]	吊杆
suspension bridge	悬索桥
sustain [səˈsteɪn]	维持；支撑，支持
sweep	蜿蜒
swell [swel]	增强；肿胀；膨胀
swelling soil	膨胀土
swelling	膨胀的，增大的
symmetrical [sɪˈmetrɪkl]	对称的
synthetic aggregate	合成骨料
tamping roller	羊足路碾，夯击式压路机
tangent [ˈtændʒənt]	【数】正切；（铁路或道路的）直线区间

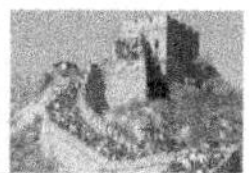

tangent grade	直坡线，直坡段
tape	卷尺
tar [tɑ:]	焦油，沥青，柏油
technical specification	技术规范
telescope [ˈtelɪskəʊp]	望远镜
temporary bridge	临时桥
temporary casing	临时护筒
tender document	投标文件
tensile strength	抗张强度
tension	【物】张力，拉力
terrain [təˈreɪn]	地形，地势；地面，地带；【地理】岩层
terrain model	地面模型
territory[ˈterətri]	领土，领地
thaw [θɔ:]	解冻，融解，回暖
the external load	外加载荷
Ministry of Housing and Urban-Rural Development	住房和城乡
the Ministry of Transport	交通运输部
the point of collapse	破坏点
the preliminary planning	初步规划
the rebar cage	钢筋笼
theodolite [θiˈɒdəlaɪt]	经纬仪
thermal [ˈθɜ:ml]	热的
thermal cracking	加热分裂（法），热裂化，热裂解，热破裂
three-dimensional	三维的
three-dimensional coordinate system	三维坐标系统
three-way interchange	三方向立体交叉道
through bridge	下承式桥
thrust [θrʌst]	推力
tilt[tɪlt]	倾斜
timber bridge	木桥
tire pressure	轮胎气压
toe [təʊ]	脚尖
toll plaza	（道路上的）收费站，收费区，收费广场
toll road	收费公路

tollway	【美】收费公路
topography[təˈpɒgrəfi]	地形学
torsion [ˈtɔːʃn]	扭转力
total station	全站仪
total station theodolite	全站仪
tow	拖拽
tower	塔，塔柱，索塔
toxic substances	有毒物质
track	小路，小道
traffic capacity	交通（容）量
traffic jam	交通堵塞
traffic load	交通载荷，行车载荷
traffic service	交通服务
traffic sign	交通标志
traffic volume	交通量
traffic wear	交通磨损
transfer	使转移
transient[ˈtrænziənt]	短暂的，临时的
transit [ˈtrænzɪt]	theodolite 经纬仪
transition	过渡，转变，变迁
translation [trænsˈleɪʃn]	平移
translational movement	平移
transmit [trænsˈmɪt]	传递，传送
transmitter	发射器
transportation engineer	交通工程师
transverse [ˈtrænzvɜːs]	横向的
trench [trentʃ]	深沟，地沟，沟渠
trim	修剪；整理
tripod [ˈtraɪpɒd]	【摄】三脚架
tropical rain forest	热带雨林
trumpet [ˈtrʌmpɪt]	喇叭
trumpet interchange	喇叭形立体交叉道
truss [trʌs]	桁架
truss bridge	桁架桥梁

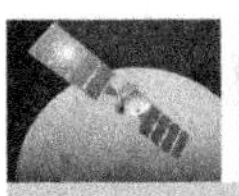

trussed arch bridge	桁架拱桥
tunnel	隧道
turbine interchange	漩涡形立体交叉道
turnkey contract	包括规划、设计和管理的施工合同，整套承包合同
turnkey[ˈtɜ:nki:]	包到底的工程，交钥匙工程
twisting moment	扭矩
two-lane road	双车道道路
two-way interchange	双方向立体交叉道
underlying	基础的；表面下的，下层的
underpass	地下通道
uneven settlement of the foundation	地基不均匀沉降
unit price	单价，分项价格
un-priced technical bid	不确定费用技术招标
unsafe [ʌnˈseɪf]	不安全的，危险的
upward	向上的
urban road	城市道路
urgent	急迫的；催促的
utility [ju:ˈtɪləti]	功用，效用
utilize	使用
U-turn ramp	回转匝道
validity [vəˈlɪdəti]	有效，合法性；效力
valley [ˈvæli]	山谷，溪谷
vandalism [ˈvændəlɪzəm]	故意破坏公共财物罪；恣意毁坏他人财产罪
variable load	可变载荷
vehicle [ˈvi:əkl]	车辆，交通工具
vehicular [viˈhɪkjələ(r)]	车的，用车辆运载的
vehicular load	行车载荷
velocity[vəˈlɒsəti]	速度
vernier ['vɜ:nɪə]	游尺，游标，游标尺
versus [ˈvɜ:səs]	与……相对；对抗
vertical [ˈvɜ:tɪkl]	adj.垂直的；n.垂直线，垂直面
vertical alignment	竖向定线，竖向线型
vertical curve	竖曲线
vertical load	竖向载荷

vessel [ˈvesl]	船只
vibrate [vaɪˈbreɪt]	（使）振动
vibration [vaɪˈbreɪʃn]	振动
vibratory compactor	振动压路机
viscous [ˈvɪskəs]	黏性的；半流体的；黏滞的
void[vɔɪd]	空的，空间
volcanic ash	火山灰
volume	体积，容积
volumetric	测定体积的
volumetric change	体积变化
warrant	保证
waterproof [ˈwɔ:təpru:f]	adj.不透水的，防水的；vt.使防水，使不透水
waterway [ˈwɔ:təweɪ]	航道
wearing course	磨耗层
wearing surface	磨耗面，磨损面
weather	风化
weight restriction	限重
weld [weld]	焊接
well-defined	定义明确的；界限清晰的
wheel load	轮载荷
whirl [wɜ:l]	旋转，回旋
wind load	风力载荷；风载荷
windward [ˈwɪndwəd]	adj.& adv.迎风的（地）；n.上风，迎风；
wingwall	翼墙
within the allowable limits	在允许的范围内
withstand [wɪðˈstænd]	经受，承受
working section	施工队
yield	屈服
yield point	屈服点

参考文献

[1] 赵永平. 路桥工程专业英语[M]. 北京：人民交通出版社，2007.

[2] 贾艳敏. 土木工程专业英语（第二版）[M]. 北京：科学出版社，2011.

[3] 李嘉. 专业英语（第三版）[M]. 北京：人民交通出版社，2012.

[4] WHIIAMS A. Design of Reinforced Concrete Structures [M]. San Jose, CA: Engineering Press, 2000.

[5] WRIGHT P H, PAQUETTE R J. Highway Engineering [M]. New York: John Wiley & Sons, 1987.

[6] BEDFORD A, FOWLER W, UECHTI K M. Statics and Mechanics of Materials [Ml. Upper Saddle River, N J: Prentice Hall, 2003.

[7] BEST R and VALENCE G D. Design and Construction [M]. Oxford, U.K.: Butterworth-Heinenmann, 2002.

[8] IAN J. Municipal Engineer — the Silver Anniversary [J]. Proceedings of ICE — Municipal Engineer, 2009, 162(2):65-68.

[9] NEIL B. Briefing Note — International Federation of Municipal Engineering [J]. Proceedings of ICE — Municipal Engineer, 2009, 163(3):1.

[10] WANG J H，KOIZUMI A，LIU X. Advancing Sustainable Urban Development in China [J]. Proceedings of ICE — Municipal Engineer, 2008, 161(1):3-10.

[11] 陈焕辉. 土木建筑英语[M]. 上海：复旦大学出版社，2007.

[12] 柴金义. 专业英语[M]. 北京：人民交通出版社，1998.

[13] 杨建华. 新概念英语互译[M]. 天津：天津大学出版社，2001.

[14] 段兵延. 土木工程专业英语（第 3 版）[M]. 武汉：武汉理工大学出版社，2018.

[15] 宋云连，崔亚楠. 道路桥梁与交通工程专业英语[M]. 北京：中国水利水电出版社，2012.

[16] 金伟良，吕清芳，潘仁泉. 东南沿海公路桥梁耐久性现状[J]. 江苏大学学报：自然科学版，2007, 28(3):4.